T0332898

Bayesian Networks

CHAPMAN & HALL/CRC
Texts in Statistical Science Series

Joseph K. Blitzstein, *Harvard University, USA*
Julian J. Faraway, *University of Bath, UK*
Martin Tanner, *Northwestern University, USA*
Jim Zidek, *University of British Columbia, Canada*

Recently Published Titles

Statistical Analysis of Financial Data
With Examples in R
James Gentle

Statistical Rethinking
A Bayesian Course with Examples in R and STAN, Second Edition
Richard McElreath

Statistical Machine Learning
A Model-Based Approach
Richard Golden

Randomization, Bootstrap and Monte Carlo Methods in Biology
Fourth Edition
Bryan F. J. Manly, Jorje A. Navarro Alberto

Principles of Uncertainty, Second Edition
Joseph B. Kadane

Beyond Multiple Linear Regression
Applied Generalized Linear Models and Multilevel Models in R
Paul Roback, Julie Legler

Bayesian Thinking in Biostatistics
Gary L. Rosner, Purushottam W. Laud, and Wesley O. Johnson

Linear Models with Python
Julian J. Faraway

Modern Data Science with R, Second Edition
Benjamin S. Baumer, Daniel T. Kaplan, and Nicholas J. Horton

Probability and Statistical Inference
From Basic Principles to Advanced Models
Miltiadis Mavrakakis and Jeremy Penzer

Bayesian Networks
With Examples in R, Second Edition
Marco Scutari and Jean-Baptiste Denis

**For more information about this series, please visit: https://www.crcpress.com/
Chapman--Hall/CRC-Texts-in-Statistical-Science/book-series/CHTEXSTASCI**

Bayesian Networks
With Examples in R
Second Edition

Marco Scutari
Jean-Baptiste Denis

CRC Press
Taylor & Francis Group
Boca Raton London New York

CRC Press is an imprint of the
Taylor & Francis Group, an **informa** business

A CHAPMAN & HALL BOOK

Second edition published 2022
by CRC Press
6000 Broken Sound Parkway NW, Suite 300, Boca Raton, FL 33487-2742

and by CRC Press
2 Park Square, Milton Park, Abingdon, Oxon, OX14 4RN

© 2022 Taylor & Francis Group, LLC

First edition published by CRC Press Taylor & Francis Group, 2014

CRC Press is an imprint of Taylor & Francis Group, LLC

Library of Congress Cataloging-in-Publication Data

ISBN: 978-0-367-36651-3 (hbk)
ISBN: 978-0-429-34743-6 (ebk)
ISBN: 978-1-032-03849-0 (pbk)

DOI: 10.1201/9780429347436

Typeset in LMR10 font
by KnowledgeWorks Global Ltd.

Visit the companion website/eResources: https://www.bnlearn.com/book-crc-2ed/

To the UK,
my home for the last decade

To my wife, Jeanie

.

Contents

Preface to the Second Edition

We have been gratified by the popularity of the first edition of *Bayesian Networks with Examples in R*, which has been well beyond our expectations. Following the publication of the first edition, the book has been translated to French by Jean-Baptiste, and a Japanese translation by Kyoritsu Shuppan is in the works (possibly due in 2021). It has been cited more than 250 times according to Google Scholar, and many readers got in touch with us over the years while using it as a reference in their own work.

This motivated us to produce a second edition containing several improvements: the result is a somewhat longer book and one we hope will answer more of the common questions that practitioners have when using Bayesian networks in their work. Firstly, we would like to thank the many readers who have reported assorted errors and typos. We gladly owned up to any mistakes we did in preparing the first edition of the book, and we have done our best not to introduce more in this new edition. We would also like to thank our editor, John Kimmel, for his support and encouragement. Secondly, we have included new material on topics chosen by popular demand:

- conditional Gaussian Bayesian networks (Chapter 3);
- dynamic Bayesian networks (Chapter 4);
- a new chapter on general Bayesian networks (Chapter 5), now using Stan instead of JAGS;
- better coverage of exact inference via junction trees, with a step-by-step example (Section 6.6.2.1); and
- how to assess the quality of a Bayesian network (Section 6.8).

The homepage for this new edition is:

https://www.bnlearn.com/book-crc-2ed/

The R code used in the book can be downloaded from there: we extracted it from the book sources when compiling them with the **knitr** package. We also provide the list of the R packages we used and their reference version, as well as any errata that we will become aware of.

Lugano, Switzerland   *Marco Scutari*
Guyancourt, France   *Jean-Baptiste Denis*
January 2021

Preface to the First Edition

Applications of Bayesian networks have multiplied in recent years, spanning such different topics as systems biology, economics, social sciences and medical informatics. Different aspects and properties of this class of models are crucial in each field: the possibility of learning causal effects from observational data in social sciences, where collecting experimental data is often not possible; the intuitive graphical representation, which provides a qualitative understanding of pathways in biological sciences; the ability to construct complex hierarchical models for phenomena that involve many interrelated components, using the most appropriate probability distribution for each of them. However, all these capabilities are built on the solid foundations provided by a small set of core definitions and properties, on which we will focus for most of the book. Handling high-dimensional data and missing values, the fine details of causal reasoning, learning under sets of additional assumptions specific to a particular field, and other advanced topics are beyond the scope of this book. They are thoroughly explored in monographs such as Nagarajan et al. (2013), Pourret et al. (2008) and Pearl (2009).

The choice of the R language is motivated, likewise, by its increasing popularity across different disciplines. Its main shortcoming is that R only provides a command-line interface, which comes with a fairly steep learning curve and is intimidating to practitioners of disciplines in which computer programming is not a core topic. However, once mastered, R provides a very versatile environment for both data analysis and the prototyping of new statistical methods. The availability of several contributed packages covering various aspects of Bayesian networks means that the reader can explore the contents of this book without reimplementing standard approaches from literature. Among these packages, we focus mainly on **bnlearn** (written by the first author, at version 3.5 at the time of this writing) to allow the reader to concentrate on studying Bayesian networks without having to first figure out the peculiarities of each package. A much better treatment of their capabilities is provided in Højsgaard et al. (2012) and in the respective documentation resources, such as vignettes and reference papers.

Bayesian Networks: With Examples in R aims to introduce the reader to Bayesian networks using a hands-on approach, through simple yet meaningful examples explored with the R software for statistical computing. Indeed, being *hands-on* is a key point of this book, in that the material strives to detail each modelling step in a simple way and with supporting R code. We know very well that a number of good books are available on this topic, and we referenced them in the "Further Reading" sections at the end of each chapter. However, we feel that the way we chose to present the material is different and that it makes this book suitable for a first introductory overview of Bayesian networks. At the same time, it may also provide a practical way to use, thanks to R, such a versatile class of models.

We hope that the book will also be useful to non-statisticians working in very different fields. Obviously, it is not possible to provide worked-out examples covering every field in which Bayesian networks are relevant. Instead, we prefer to give a clear understanding of the general approach and of the steps it involves. Therefore, we explore a limited number of examples in great depth, considering that experts will be able to reinterpret them in the respective fields. We start from the simplest notions, gradually increasing complexity in later chapters. We also distinguish the probabilistic models from their estimation with data sets: when the separation is not clear, confusion is apparent when performing inference.

Bayesian Networks: With Examples in R is suitable for teaching in a semester or half-semester course, possibly integrating other books. More advanced theoretical material and the analysis of two real-world data sets are included in the second half of the book for further understanding of Bayesian networks. The book is targeted at the level of a M.Sc. or Ph.D. course, depending on the background of the student. In the case of disciplines such as mathematics, statistics and computer science the book is suitable for M.Sc. courses, while for life and social sciences the lack of a strong grounding in probability theory may make the book more suitable for a Ph.D. course. In the former, the reader may prefer to first review the second half of the book, to grasp the theoretical aspects of Bayesian networks before applying them; while in the latter he can get a hang of what Bayesian networks are about before investing time in studying their underpinnings. Introductory material on probability, statistics and graph theory is included in the appendixes. Furthermore, the solutions to the exercises are included in the book for the convenience of the reader. The real-world examples in the last chapter will motivate students by showing current applications in the literature. Introductory examples in earlier chapters are more varied in topic, to present simple applications in different contexts.

The skills required to understand the material are mostly at the level of a B.Sc. graduate. Nevertheless, a few topics are based on more specialised concepts whose illustration is beyond the scope of this book. The basics of R programming are not covered in the book, either, because of the availability of accessible and thorough references such as Venables and Ripley (2002), Spector (2009) and Crawley (2013). Basic graph and probability theory are covered in the appendixes for easy reference. Pointers to literature are provided at the end of each chapter, and supporting material will be available online from www.bnlearn.com.

The book is organised as follows. Discrete Bayesian networks are described first (Chapter 1), followed by Gaussian Bayesian networks (Chapter 2).

Hybrid networks (which include arbitrary random variables, and typically mix continuous and discrete ones) are covered in Chapter 3. These chapters explain the whole process of Bayesian network modelling, from structure learning to parameter learning to inference. All steps are illustrated with R code. A concise but rigorous treatment of the fundamentals of Bayesian networks is given in Chapter 6, and includes a brief introduction to causal Bayesian networks. For completeness, we also provide an overview of the available software in Chapter 7, both in R and other software packages. Subsequently, two real-world examples are analysed in Chapter 8. The first replicates the study in the landmark causal protein-signalling network paper published in *Science* by Sachs et al. (2005). The second investigates possible graphical modelling approaches in predicting the contributions of fat, lean and bone to the composition of different body parts.

Last but not least, we are immensely grateful to friends and colleagues who helped us in planning and writing this book, and its French version *Résaux Bayésiens avec R: élaboration, manipulation et utilisation en modélisation appliquée*. We are also grateful to John Kimmel of Taylor & Francis for his dedication in improving this book and organising draft reviews. We hope not to have unduly raised his stress levels, as we did our best to incorporate the reviewers' feedback and we even submitted the final manuscript on time. Likewise, we thank the people at EDP Sciences for their interest in publishing a book on this topic: they originally asked the second author to write a book in French. He was not confident enough to write a book alone and looked for a co-author, thus starting the collaboration with the first author and a wonderful exchange of ideas. The latter, not being very proficient in the French language, prepared the English draft from which this Chapman & Hall book originates. The French version is also planned to be in print by the end of this year.

London, United Kingdom   *Marco Scutari*
Jouy-en-Josas, France   *Jean-Baptiste Denis*
March 2014

1

The Discrete Case: Multinomial Bayesian Networks

In this chapter we will introduce the fundamental ideas behind Bayesian networks (BNs) and their interpretation using a hypothetical survey on the usage of different means of transport. We will focus on modelling discrete data, leaving continuous data to Chapter 2 and more complex data types to Chapters 3 and 5.

1.1 Introductory Example: Train-Use Survey

Consider a simple, hypothetical survey whose aim is to investigate the usage patterns of different means of transport, with a focus on cars and trains. Such surveys are used to assess customer satisfaction across different social groups, to evaluate public policies and to improve urban planning. Some real-world examples can be found, for instance, in Kenett et al. (2012).

In our current example we will examine, for each individual, the following six discrete variables (labels used in computations and figures are reported in parenthesis):

- **Age** (A): the age, recorded as *young* (young) for individuals below 30 years old, *adult* (adult) for individuals between 30 and 60 years old, and *old* (old) for people older than 60.

- **Sex** (S): the biological sex, recorded as *male* (M) or *female* (F).

- **Education** (E): the highest level of education or training successfully completed, recorded as *up to high school* (high) or *university degree* (uni).

- **Occupation** (O): whether the individual is an *employee* (emp) or a *self-employed* (self) worker.

- **Residence** (R): the size of the city the individual lives in, recorded as either *small* (small) or *big* (big).

- **Travel** (T): the means of transport favoured by the individual, recorded either as *car* (car), *train* (train) or *other* (other).

In the scope of this survey, each variable falls into one of three groups. Age and Sex are *demographic indicators*. In other words, they are intrinsic characteristics of the individual; they may result in different patterns of behaviour but are not influenced by the individual himself. On the other hand, the opposite is true for Education, Occupation and Residence. These variables are *socioeconomic indicators* and describe the individual's position in society. Therefore, they provide a rough description of the individual's expected lifestyle; for example, they may characterise his spending habits and his work schedule. The last variable, Travel, is the *target* of the survey, the quantity of interest whose behaviour is under investigation.

1.2 Graphical Representation

The nature of the variables recorded in the survey, and more in general of the three categories they belong to, suggests how they may be related with each other. Some of these relationships will be *direct*, while others will be mediated by one or more variables (*indirect*).

Both kinds of relationships can be represented effectively and intuitively by means of a *directed graph*, which is one of the two fundamental entities characterising a BN. Each *node* in the graph corresponds to one of the variables in the survey. In fact, they are usually referred to interchangeably in the literature. Therefore, the graph produced from this example will contain six nodes, labelled after the variables (A, S, E, O, R and T). Direct dependence relationships are represented as *arcs* between pairs of variables (*e.g.*, A → E means that E depends on A). The node at the tail of the arc is called the *parent*, while that at the head (where the arrow is) is called the *child*. Indirect dependence relationships are not explicitly represented. However, they can be read from the graph as sequences of arcs leading from one variable to the other through one or more mediating variables (*e.g.*, the combination of A → E and E → R means that R depends on A through E). Such sequences of arcs are said to form a *path* leading from one variable to the other; these two variables must be distinct. Paths of the form A → ... → A, which are known as *cycles*, are not allowed in the graph. For this reason, the graphs used in BNs are called *directed acyclic graphs* (DAGs).

Note, however, that some caution must be exercised in interpreting both direct and indirect dependencies. The presence of arrows or arcs seems to imply, at an intuitive level, that for each arc one variable should be interpreted as a *cause* and the other as an *effect* (*e.g.*, A → E means that A causes E). This interpretation, which is called *causal*, is difficult to justify in most situations: for this reason, in general we speak about dependence relationships instead of causal effects. The assumptions required for causal BN modelling will be discussed in Section 6.7.

To create and manipulate DAGs in the context of BNs, we will mainly use the **bnlearn** package (short for "**B**ayesian **n**etwork **learn**ing").

```
library(bnlearn)
```

As a first step, we create a DAG with one node for each variable in the survey and no arcs.

```
dag <- empty.graph(nodes = c("A", "S", "E", "O", "R", "T"))
```

Such a DAG is usually called an *empty graph*, because it has an empty arc set. The DAG is stored in an object of class bn, which looks as follows when printed.

```
dag

  Random/Generated Bayesian network

  model:
   [A][S][E][O][R][T]
  nodes:                               6
  arcs:                                0
    undirected arcs:                   0
    directed arcs:                     0
  average markov blanket size:         0.00
  average neighbourhood size:          0.00
  average branching factor:            0.00

  generation algorithm:                Empty
```

Now we can start adding the arcs that encode the direct dependencies between the variables in the survey. As we said in the previous section, Age and Sex are not influenced by any of the other variables. Therefore, there are no arcs pointing to either variable. On the other hand, both Age and Sex have a direct influence on Education. It is well known, for instance, that the number of people attending universities has increased over the years. As a consequence, younger people are more likely to have a university degree than older people.

```
dag <- set.arc(dag, from = "A", to = "E")
```

Similarly, Sex also influences Education; the gender gap in university applications has been widening for many years, with women outnumbering and outperforming men.

```
dag <- set.arc(dag, from = "S", to = "E")
```

In turn, Education strongly influences both Occupation and Residence. Clearly, higher education levels help in accessing more prestigious professions. In addition, people often move to attend a particular university or to find a job that matches the skills they acquired in their studies.

```
dag <- set.arc(dag, from = "E", to = "O")
dag <- set.arc(dag, from = "E", to = "R")
```

Finally, the preferred means of transport are directly influenced by both Occupation and Residence. For the former, the reason is that some jobs require periodic long-distance trips, while others require more frequent trips but on shorter distances. For the latter, the reason is that both commute time and distance are deciding factors in choosing between travelling by car or by train.

```
dag <- set.arc(dag, from = "O", to = "T")
dag <- set.arc(dag, from = "R", to = "T")
```

Now that we have added all the arcs, the DAG in the dag object encodes the desired direct dependencies. Its structure is shown in Figure 1.1, and can be read from the model formula generated from the dag object itself.

```
dag

  Random/Generated Bayesian network

  model:
   [A][S][E|A:S][O|E][R|E][T|O:R]
  nodes:                                    6
  arcs:                                     6
    undirected arcs:                        0
    directed arcs:                          6
  average markov blanket size:              2.67
  average neighbourhood size:               2.00
  average branching factor:                 1.00

  generation algorithm:                     Empty
```

Direct dependencies are listed for each variable, denoted by a bar (|) and separated by semicolons (:). For example, [E|A:S] means that A → E and S → E; while [A] means that there is no arc pointing towards A. This representation of the graph structure is designed to recall a product of conditional probabilities, for reasons that will be clear in the next section, and can be produced with the modelstring function.

```
modelstring(dag)
 [1] "[A][S][E|A:S][O|E][R|E][T|O:R]"
```

bnlearn provides many other functions to investigate and manipulate bn objects. For a comprehensive overview, we refer the reader to the documentation included in the package. Two basic examples are nodes and arcs.

```
nodes(dag)
 [1] "A" "S" "E" "O" "R" "T"
```

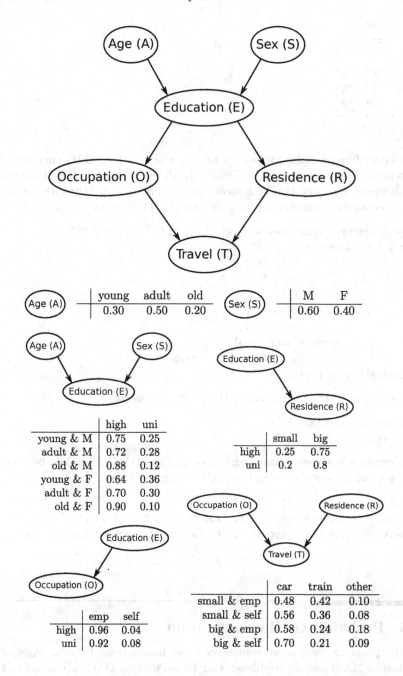

Figure 1.1
DAG representing the dependence relationships linking the variables recorded in the survey: Age (A), Sex (S), Education (E), Occupation (O), Residence (R) and Travel (T). The corresponding conditional probability tables are reported below.

```
arcs(dag)
     from to
[1,] "A"  "E"
[2,] "S"  "E"
[3,] "E"  "O"
[4,] "E"  "R"
[5,] "O"  "T"
[6,] "R"  "T"
```

The latter function also provides a way to add arcs to a DAG that is faster than setting them one at a time. Obviously, the approach we used above is too cumbersome for large DAGs. Instead, we can create a matrix with the same structure as that returned by arcs and set the whole arc set at once.

```
dag2 <- empty.graph(nodes = c("A", "S", "E", "O", "R", "T"))
arc.set <- matrix(c("A", "E",
                    "S", "E",
                    "E", "O",
                    "E", "R",
                    "O", "T",
                    "R", "T"),
             byrow = TRUE, ncol = 2,
             dimnames = list(NULL, c("from", "to")))
arcs(dag2) <- arc.set
```

The resulting DAG is identical to the previous one, dag.

```
all.equal(dag, dag2)
 [1] TRUE
```

Furthermore, both approaches guarantee that the DAG will indeed be acyclic; trying to introduce a cycle in the DAG returns an error.

```
set.arc(dag, from = "T", to = "E")
 Error in arc.operations(x = x, from = from, to = to, op = "set",
 check.cycles = check.cycles, : the resulting graph contains cycles.
```

1.3 Probabilistic Representation

In the previous section we represented the interactions between Age, Sex, Education, Occupation, Residence and Travel using a DAG. To complete the BN modelling the survey, we will now specify a joint probability distribution over these variables. All of them are discrete and defined on a set of non-ordered states (called *levels* in R).

```
A.lv <- c("young", "adult", "old")
S.lv <- c("M", "F")
E.lv <- c("high", "uni")
O.lv <- c("emp", "self")
R.lv <- c("small", "big")
T.lv <- c("car", "train", "other")
```

Therefore, the natural choice for the joint probability distribution is a multinomial distribution, assigning a probability to each combination of states of the variables in the survey. In the context of BNs, this joint distribution is called the *global distribution*.

However, using the global distribution directly is difficult: even for small problems, such as that we are considering, the number of parameters involved is very high. In the case of this survey, the parameter set includes the 143 probabilities corresponding to the combinations of the levels of all the variables. Fortunately, we can use the information encoded in the DAG to break down the global distribution into a set of smaller *local distributions*, one for each variable. Recall that arcs represent direct dependencies: if there is an arc from one variable to another, the latter depends on the former. In other words, variables that are not linked by an arc are *conditionally independent*. As a result, we can factorise the global distribution as follows:

$$\Pr(A, S, E, O, R, T) = \Pr(A)\Pr(S)\Pr(E \mid A, S)\Pr(O \mid E)\Pr(R \mid E)\Pr(T \mid O, R). \quad (1.1)$$

Equation (1.1) provides a formal definition of how the dependencies encoded in the DAG *map* into the probability space via conditional independence relationships. The absence of cycles in the DAG ensures that the factorisation is well defined. Each variable depends only on its parents; its distribution is univariate and has a (comparatively) small number of parameters. The set of all the local distributions has, overall, fewer parameters than the global distribution. The latter represents a more general model than the former, because it does not make any assumption on the dependencies between the variables. In other words, the factorisation in Equation (1.1) defines a *nested model* or a *submodel* of the global distribution.

In our survey, Age and Sex are modelled by simple, unidimensional probability tables (they have no parent).

```
A.prob <- array(c(0.30, 0.50, 0.20), dim = 3, dimnames = list(A = A.lv))
A.prob
A
young adult   old
  0.3   0.5   0.2
S.prob <- array(c(0.60, 0.40), dim = 2, dimnames = list(S = S.lv))
S.prob
S
  M   F
0.6 0.4
```

Occupation and Residence, which depend on Education, are modelled by two-dimensional conditional probability tables. Each column corresponds to one level of the parent and holds the distribution of the variable conditional on that particular level. As a result, probabilities sum up to 1 within each column.

```
O.prob <- array(c(0.96, 0.04, 0.92, 0.08), dim = c(2, 2),
            dimnames = list(O = O.lv, E = E.lv))
O.prob
      E
O       high  uni
  emp  0.96 0.92
  self 0.04 0.08
R.prob <- array(c(0.25, 0.75, 0.20, 0.80), dim = c(2, 2),
            dimnames = list(R = R.lv, E = E.lv))
R.prob
      E
R        high uni
  small 0.25 0.2
  big   0.75 0.8
```

For these one- and two-dimensional distributions, we can also use the matrix function to create the (conditional) probability tables. The syntax is almost identical to that of array; the difference is that only one dimension (either the number of rows, nrow, or the number of columns, ncol) must be specified.

```
R.prob <- matrix(c(0.25, 0.75, 0.20, 0.80), ncol = 2,
          dimnames = list(R = R.lv, E = E.lv))
```

Finally, Education and Travel are modelled as three-dimensional tables, since they have two parents each (Age and Sex for Education, Occupation and Residence for Travel). Each column corresponds to one combination of the levels of the parents and holds the distribution of the variable conditional on that particular combination.

```
E.prob <- array(c(0.75, 0.25, 0.72, 0.28, 0.88, 0.12, 0.64, 0.36, 0.70,
                0.30, 0.90, 0.10), dim = c(2, 3, 2),
          dimnames = list(E = E.lv, A = A.lv, S = S.lv))

T.prob <- array(c(0.48, 0.42, 0.10, 0.56, 0.36, 0.08, 0.58, 0.24, 0.18,
                0.70, 0.21, 0.09), dim = c(3, 2, 2),
          dimnames = list(T = T.lv, O = O.lv, R = R.lv))
```

Overall, the local distributions we defined above have just 21 parameters, compared to the 143 of the global distribution. Furthermore, local distributions can be handled independently from each other and have at most 8 parameters each. This reduction in dimension is a fundamental property of BNs and makes their application feasible for high-dimensional problems.

Now that we have defined both the DAG and the local distribution corresponding to each variable, we can combine them to form a fully-specified BN. For didactic purposes, we recreate the DAG using the model formula interface provided by modelstring, whose syntax is almost identical to Equation (1.1). The nodes and the parents of each node can be listed in any order, thus allowing us to follow the logical structure of the network in writing the formula.

```
dag3 <- model2network("[A][S][E|A:S][O|E][R|E][T|O:R]")
```

The resulting DAG is identical to that we created in the previous section, as shown below.

```
all.equal(dag, dag3)
  [1] TRUE
```

Then we combine the DAG we stored in dag and a list containing the local distributions, which we will call cpt, into an object of class bn.fit called bn.

```
cpt <- list(A = A.prob, S = S.prob, E = E.prob, O = O.prob, R = R.prob,
        T = T.prob)
bn <- custom.fit(dag, cpt)
```

The number of parameters of the BN can be computed with the nparams function and is indeed equal to 21, as expected from the parameter sets of the local distributions.

```
nparams(bn)
  [1] 21
```

Objects of class bn.fit are used to describe BNs in **bnlearn**. They include information about both the DAG (such as the parents and the children of each node) and the local distributions (their parameters). For most practical purposes, they can be used as if they were objects of class bn when investigating graphical properties. So, for example,

```
arcs(bn)
      from to
  [1,] "A"  "E"
  [2,] "S"  "E"
  [3,] "E"  "O"
  [4,] "E"  "R"
  [5,] "O"  "T"
  [6,] "R"  "T"
```

and the same holds for other functions such as nodes, parents and children. Furthermore, the conditional probability tables can either be printed directly from the bn.fit object

```
bn$R

  Parameters of node R (multinomial distribution)

Conditional probability table:

      E
R       high  uni
  small 0.25 0.20
  big   0.75 0.80
```

or extracted for later use with the coef function as follows.

```
R.cpt <- coef(bn$R)
```

Just typing

```
bn
```

prints all the conditional probability tables in the BN.

1.4 Estimating the Parameters: Conditional Probability Tables

For the hypothetical survey described in this chapter, we have assumed to know both the DAG and the parameters of the local distributions defining the BN. In this scenario, BNs are used as *expert systems* because they formalise the knowledge possessed by one or more experts in the relevant fields. However, in most cases the parameters of the local distributions will be estimated (or *learned*) from an observed sample. Typically, the data will be stored in a text file that we can import with read.table,

```
survey <- read.table("survey.txt", header = TRUE, colClasses = "factor")
```

with one variable per column (labelled in the first row) and one observation per line.

```
head(survey)
        A     R    E   O S     T
1 adult   big high emp F   car
2 adult small  uni emp M   car
3 adult   big  uni emp F train
4 adult   big high emp M   car
5 adult   big high emp M   car
6 adult small high emp F train
```

In the case of this survey, and of discrete BNs in general, the parameters to estimate are the conditional probabilities in the local distributions. They can be estimated, for example, with the corresponding empirical frequencies in the data set, *e.g.*,

$$\widehat{\Pr}(\mathtt{O} = \mathtt{emp} \mid \mathtt{E} = \mathtt{high}) = \frac{\widehat{\Pr}(\mathtt{O} = \mathtt{emp}, \mathtt{E} = \mathtt{high})}{\widehat{\Pr}(\mathtt{E} = \mathtt{high})} =$$

$$= \frac{\text{number of observations for which } \mathtt{O} = \mathtt{emp} \text{ and } \mathtt{E} = \mathtt{high}}{\text{number of observations for which } \mathtt{E} = \mathtt{high}}. \quad (1.2)$$

This yields the classic *frequentist* and *maximum likelihood* estimates. In **bnlearn**, we can compute them with the bn.fit function. bn.fit complements the custom.fit function we used in the previous section; the latter constructs a BN using a set of *custom* parameters specified by the user, while the former estimates them from the data.

```
bn.mle <- bn.fit(dag, data = survey, method = "mle")
```

Similarly to custom.fit, bn.fit returns an object of class bn.fit. The method argument determines which estimator will be used; in this case, "mle" for the maximum likelihood estimator. Again, the structure of the network is assumed to be known and is passed to the function via the dag object. For didactic purposes, we can also compute the same estimates manually

```
prop.table(table(survey[, c("O", "E")]), margin = 2)
        E
O        high    uni
   emp  0.9808  0.9259
   self 0.0192  0.0741
```

and verify that we get the same result as bn.fit.

```
bn.mle$O

  Parameters of node O (multinomial distribution)

Conditional probability table:

        E
O        high    uni
   emp  0.9808  0.9259
   self 0.0192  0.0741
```

As an alternative, we can also estimate the same conditional probabilities in a Bayesian setting, using their posterior distributions. An overview of the underlying probability theory and the distributions relevant for BNs is provided in Appendixes B.3, B.4 and B.5. In this case, the method argument of bn.fit must be set to "bayes".

```
bn.bayes <- bn.fit(dag, data = survey, method = "bayes", iss = 10)
```

The estimated posterior probabilities are computed from a uniform prior over each conditional probability table. The iss optional argument, whose name stands for *imaginary sample size* (also known as *equivalent sample size*), determines how much weight is assigned to the prior distribution compared to the data when computing the posterior. The weight is specified as the size of an imaginary sample supporting the prior distribution. Its value is divided by the number of cells in the conditional probability table (because the prior is flat) and used to compute the posterior estimate as a weighted mean with the empirical frequencies. So, for example, suppose we have a sample of size n, which we can compute as nrow(survey). If we let

$$\hat{p}_{\text{emp,high}} = \frac{\text{number of observations for which O = emp and E = high}}{n} \quad (1.3)$$

$$\hat{p}_{\text{high}} = \frac{\text{number of observations for which E = high}}{n} \quad (1.4)$$

and we denote the corresponding prior probabilities as

$$\pi_{\text{emp,high}} = \frac{1}{\text{nO} \times \text{nE}} \quad \text{and} \quad \pi_{\text{high}} = \frac{\text{nO}}{\text{nO} \times \text{nE}} \quad (1.5)$$

where nO = nlevels(bn.bayes\$O) and nE = nlevels(bn.bayes\$E), we have that

$$\widehat{\Pr}(\text{O = emp, E = high}) = \frac{\text{iss}}{n + \text{iss}} \pi_{\text{emp,high}} + \frac{n}{n + \text{iss}} \hat{p}_{\text{emp,high}} \quad (1.6)$$

$$\widehat{\Pr}(\text{E = high}) = \frac{\text{iss}}{n + \text{iss}} \pi_{\text{high}} + \frac{n}{n + \text{iss}} \hat{p}_{\text{high}} \quad (1.7)$$

and therefore that

$$\widehat{\Pr}(\text{O = emp} \mid \text{E = high}) = \frac{\widehat{\Pr}(\text{O = emp, E = high})}{\widehat{\Pr}(\text{E = high})}. \quad (1.8)$$

The value of iss is typically chosen to be small, usually between 1 and 10, to allow the prior distribution to be easily dominated by the data. Such small values result in conditional probabilities that are smoother but still close to the empirical frequencies (*i.e.*, $\hat{p}_{\text{emp,high}}$) they are computed from.

```
bn.bayes$O

  Parameters of node O (multinomial distribution)


  Conditional probability table:


       E
  O        high    uni
    emp  0.9743 0.9107
    self 0.0257 0.0893
```

As we can see from the conditional probability table above, all the posterior estimates are farther from both 0 and 1 than the corresponding maximum likelihood estimates due to the influence of the prior distribution. This is desirable for several reasons. First of all, this ensures that the regularity conditions of model estimation and inference methods are fulfilled. In particular, it is not possible to obtain sparse conditional probability tables (with many zero cells) even from small data sets. Furthermore, posterior estimates are more robust than maximum likelihood estimates and result in BNs with better predictive power.

Increasing the value of iss makes the posterior distribution more and more flat, pushing it towards the uniform distribution used as the prior. As shown in Figure 1.2, for large values of iss the conditional posterior distributions for $\Pr(O \mid E = \text{high})$ and $\Pr(O \mid E = \text{uni})$ assign a probability of approximately 0.5 to both self and emp. This trend is already apparent if we compare the conditional probabilities obtained for iss = 10 with those for iss = 20, reported below.

```
bn.bayes <- bn.fit(dag, data = survey, method = "bayes", iss = 20)
bn.bayes$O


  Parameters of node O (multinomial distribution)


Conditional probability table:

       E
O        high   uni
  emp   0.968 0.897
  self  0.032 0.103
```

1.5 Learning the DAG Structure: Tests and Scores

In the previous sections we have assumed that the DAG underlying the BN is known. In other words, we rely on prior knowledge on the phenomenon we are modelling to decide which arcs are present in the graph and which are not. However, this is not always possible or desired; the structure of the DAG itself may be the object of our investigation. It is common in genetics and systems biology, for instance, to reconstruct the molecular pathways and networks underlying complex diseases and metabolic processes. An outstanding example of this kind of study can be found in Sachs et al. (2005) and will be explored in Chapter 8. In the context of social sciences, the structure of the DAG may identify which nodes are directly related to the target of the analysis and may therefore be used to improve the process of policy making. For instance, the

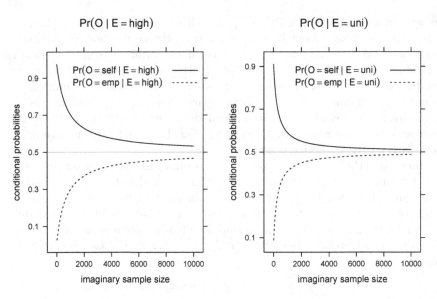

Figure 1.2
Conditional probability distributions for O given both possible values of E, that is, $\Pr(\text{O} \mid \text{E} = \text{high})$ and $\Pr(\text{O} \mid \text{E} = \text{uni})$, converge to uniform distributions as the imaginary sample size increases.

DAG of the survey we are using as an example suggests that train fares should be adjusted (to maximise profit) on the basis of Occupation and Residence alone.

Learning the DAG of a BN is a complex task, for two reasons. First, the space of the possible DAGs is very big; the number of DAGs increases super-exponentially as the number of nodes grows. As a result, only a small fraction of its elements can be investigated in a reasonable time. Furthermore, this space is very different from real spaces (*e.g.*, $\mathbb{R}$, $\mathbb{R}^2$, $\mathbb{R}^3$, etc.) in that it is not continuous and has a finite number of elements. Therefore, ad-hoc algorithms are required to explore it. We will investigate the algorithms proposed for this task and their theoretical foundations in Section 6.5. For the moment, we will limit our attention to the two classes of statistical criteria used by those algorithms to evaluate DAGs: *conditional independence tests* and *network scores*.

1.5.1 Conditional Independence Tests

Conditional independence tests focus on the presence of individual arcs. Since each arc encodes a probabilistic dependence, conditional independence tests can be used to assess whether that probabilistic dependence is supported by the data. If the null hypothesis (of conditional independence) is rejected, the arc can be considered for inclusion in the DAG. For instance, consider adding an arc from Education to Travel (E → T) to the DAG shown in Figure 1.1. The null hypothesis is that Travel is probabilistically independent ($\perp\!\!\!\perp_P$) from Education conditional on its parents, *i.e.*,

$$H_0 : \mathsf{T} \perp\!\!\!\perp_P \mathsf{E} \mid \{\mathsf{O}, \mathsf{R}\}, \tag{1.9}$$

and the alternative hypothesis is that

$$H_1 : \mathsf{T} \not\!\perp\!\!\!\perp_P \mathsf{E} \mid \{\mathsf{O}, \mathsf{R}\}. \tag{1.10}$$

We can test this null hypothesis by adapting either the *log-likelihood ratio* G^2 or *Pearson's* X^2 to test for conditional independence instead of marginal independence. For G^2, the test statistic assumes the form

$$G^2(\mathsf{T}, \mathsf{E} \mid \mathsf{O}, \mathsf{R}) = \sum_{t \in \mathsf{T}} \sum_{e \in \mathsf{E}} \sum_{k \in \mathsf{O} \times \mathsf{R}} n_{tek} \log \frac{n_{tek} n_{++k}}{n_{t+k} n_{+ek}}, \tag{1.11}$$

where we denote the categories of Travel with $t \in \mathsf{T}$, the categories of Education with $e \in \mathsf{E}$, and the configurations of Occupation and Residence with $k \in \mathsf{O} \times \mathsf{R}$. Hence, n_{tek} is the number of observations for the combination of a category t of Travel, a category e of Education and a category k of $\mathsf{O} \times \mathsf{R}$. The use of a "+" subscript denotes the sum over an index, as in the classic book from Agresti (2013), and is used to indicate the marginal counts for the remaining variables. So, for example, n_{t+k} is the number of observations for t and k obtained by summing over all the categories of Education. For Pearson's X^2, using the same notation we have that

$$X^2(\mathsf{T}, \mathsf{E} \mid \mathsf{O}, \mathsf{R}) = \sum_{t \in \mathsf{T}} \sum_{e \in \mathsf{E}} \sum_{k \in \mathsf{O} \times \mathsf{R}} \frac{(n_{tek} - m_{tek})^2}{m_{tek}}, \quad \text{where} \quad m_{tek} = \frac{n_{t+k} n_{+ek}}{n_{++k}}.$$
$$\tag{1.12}$$

Both tests have an asymptotic χ^2 distribution under the null hypothesis, in this case with

```
(nlevels(survey[, "T"]) - 1) * (nlevels(survey[, "E"]) - 1) *
  (nlevels(survey[, "O"]) * nlevels(survey[, "R"]))
[1] 8
```

degrees of freedom. Conditional independence results in small values of G^2 and X^2; conversely, the null hypothesis is rejected for large values of the test statistics, which increase with the strength of the conditional dependence between the variables.

The `ci.test` function from **bnlearn** implements both G^2 and X^2, in addition to other tests which we will cover in Section 6.5.1.1. The G^2 test, which is equivalent to the *mutual information* test from information theory, is used when `test = "mi"`.

```
ci.test("T", "E", c("O", "R"), test = "mi", data = survey)

          Mutual Information (disc.)

  data:  T ~ E | O + R
  mi = 9.88, df = 8, p-value = 0.27
  alternative hypothesis: true value is greater than 0
```

Pearson's X^2 test is used when test = "x2".

```
ci.test("T", "E", c("O", "R"), test = "x2", data = survey)

          Pearson's X^2

  data:  T ~ E | O + R
  x2 = 8.24, df = 8, p-value = 0.41
  alternative hypothesis: true value is greater than 0
```

Both tests return very large p-values, indicating that the dependence relationship encoded by $E \rightarrow T$ is not significant given the current DAG structure.

We can test in a similar way whether one of the arcs in the DAG should be removed because the dependence relationship it encodes is not supported by the data. So, for example, we can remove $O \rightarrow T$ by testing

$$H_0 : T \perp\!\!\!\perp_P O \mid R \qquad \text{versus} \qquad H_1 : T \not\perp\!\!\!\perp_P O \mid R \qquad (1.13)$$

as follows.

```
ci.test("T", "O", "R", test = "x2", data = survey)

          Pearson's X^2

  data:  T ~ O | R
  x2 = 3.8, df = 4, p-value = 0.43
  alternative hypothesis: true value is greater than 0
```

Again, we find that $O \rightarrow T$ is not significant.

The task of testing each arc in turn for significance can be automated using the arc.strength function and specifying the test label with the criterion argument.

```
arc.strength(dag, data = survey, criterion = "x2")
    from to strength
1    A  E  0.00098
2    S  E  0.00125
3    E  O  0.00264
4    E  R  0.00056
5    O  T  0.43391
6    R  T  0.00136
```

arc.strength is designed to measure the strength of the probabilistic dependence corresponding to each arc by removing that particular arc from the graph and quantifying the change with some probabilistic criterion. Possible choices are a conditional independence test (in the example above) or a network score (in the next section). In the case of conditional independence tests, the value of the criterion argument is the same as that of the test argument in ci.test, and the test is for the to node to be independent from the from node conditional on the remaining parents of to. The reported strength is the resulting p-value. What we see from the output above is that all arcs with the exception of O → T have p-values smaller than 0.05 and are well supported by the data.

1.5.2 Network Scores

Unlike conditional independence tests, network scores focus on the DAG as a whole; they are goodness-of-fit statistics measuring how well the DAG mirrors the dependence structure of the data. Again, several scores are in common use. One of them is the *Bayesian Information criterion* (BIC), which for our survey BN takes the form

$$
\text{BIC} = \log \widehat{\text{Pr}}(\mathsf{A}, \mathsf{S}, \mathsf{E}, \mathsf{O}, \mathsf{R}, \mathsf{T}) - \frac{d}{2} \log n =
$$

$$
= \left[\log \widehat{\text{Pr}}(\mathsf{A}) - \frac{d_\mathsf{A}}{2} \log n \right] + \left[\log \widehat{\text{Pr}}(\mathsf{S}) - \frac{d_\mathsf{S}}{2} \log n \right] +
$$

$$
+ \left[\log \widehat{\text{Pr}}(\mathsf{E} \mid \mathsf{A}, \mathsf{S}) - \frac{d_\mathsf{E}}{2} \log n \right] + \left[\log \widehat{\text{Pr}}(\mathsf{O} \mid \mathsf{E}) - \frac{d_\mathsf{O}}{2} \log n \right] +
$$

$$
+ \left[\log \widehat{\text{Pr}}(\mathsf{R} \mid \mathsf{E}) - \frac{d_\mathsf{R}}{2} \log n \right] + \left[\log \widehat{\text{Pr}}(\mathsf{T} \mid \mathsf{O}, \mathsf{R}) - \frac{d_\mathsf{T}}{2} \log n \right] \quad (1.14)
$$

where n is the sample size, d is the number of parameters of the whole network (*i.e.*, 21) and d_A, d_S, d_E, d_O, d_R and d_T are the numbers of parameters associated with each node. The decomposition in Equation (1.1) makes it easy to compute BIC from the local distributions. Another score commonly used in the literature is the *Bayesian Dirichlet equivalent uniform* (BDeu) posterior probability of the DAG associated with a uniform prior over both the space of the DAGs and of the parameters; its general form is given in Section 6.5. It is often denoted simply as BDe. Both BIC and BDe assign higher scores to DAGs that fit the data better.

Both scores can be computed in **bnlearn** using the score function; BIC is computed when type = "bic", and log BDe when type = "bde".

```
score(dag, data = survey, type = "bic")
[1] -2012.69
score(dag, data = survey, type = "bde", iss = 10)
[1] -1998.28
```

Note that the iss argument for BDe is the same imaginary sample size we introduced when computing posterior estimates of the BN's parameters in Section 1.4. As before, it can be interpreted as the weight assigned to the (flat) prior distribution in terms of the size of an imaginary sample. For small values of iss or large sample sizes, log BDe and BIC scores yield similar values.

```
score(dag, data = survey, type = "bde", iss = 1)
 [1] -2015.65
```

Using either of these scores it is possible to compare different DAGs and investigate which fits the data better. For instance, we can consider once more whether the DAG from Figure 1.1 fits the survey data better before or after adding the arc E → T.

```
dag4 <- set.arc(dag, from = "E", to = "T")
nparams(dag4, survey)
 [1] 29
score(dag4, data = survey, type = "bic")
 [1] -2032.6
```

Again, adding E → T is not beneficial, as the increase in $\log \widehat{\Pr}(A, S, E, O, R, T)$ is not sufficient to offset the heavier penalty from the additional parameters. The score for dag4 (-2032.6) is lower than that of dag3 (-2012.69).

Scores can also be used to compare completely different networks, unlike conditional independence tests. We can even generate a DAG at random with random.graph and compare it to the previous DAGs through its score.

```
rnd <- random.graph(nodes = c("A", "S", "E", "O", "R", "T"))
modelstring(rnd)
 [1] "[A][S|A][E|A:S][O|S:E][R|S:E][T|S:E]"
score(rnd, data = survey, type = "bic")
 [1] -2034.99
```

As expected, rnd is worse than dag and even dag4; after all, neither data nor common sense are used to select its structure! Learning the DAG from survey yields a much better network. There are several algorithms that tackle this problem by searching for the DAG that maximises a given network score; some will be illustrated in Section 6.5.1.2. A simple one is *hill-climbing*: starting from a DAG with no arcs, it adds, removes and reverses one arc at a time and picks the change that increases the network score the most. It is implemented in the hc function, which in its most basic form takes the data (survey) as the only argument and defaults to the BIC score.

```
learned <- hc(survey)
modelstring(learned)
 [1] "[R][E|R][T|R][A|E][O|E][S|E]"
score(learned, data = survey, type = "bic")
 [1] -1998.43
```

We can specify other scores with the `score` argument: for example, we can change the default `score` = "bic" to `score` = "bde".

```
learned2 <- hc(survey, score = "bde")
```

Unsurprisingly, removing any arc from `learned` decreases its BIC score. We can confirm this conveniently using `arc.strength`, which reports the change in the score caused by an arc removal as the arc's `strength` when `criterion` is a network score.

```
arc.strength(learned, data = survey, criterion = "bic")
  from to strength
1    R  E   -3.390
2    E  S   -2.726
3    R  T   -1.848
4    E  A   -1.720
5    E  O   -0.827
```

This is not true for `dag`, suggesting that not all the dependencies it encodes can be learned correctly from `survey`.

```
arc.strength(dag, data = survey, criterion = "bic")
  from to strength
1    A  E    2.489
2    S  E    1.482
3    E  O   -0.827
4    E  R   -3.390
5    O  T   10.046
6    R  T    2.973
```

In particular, removing O → T causes a marked increase in the BIC score, which is consistent with the high p-value we observed for this arc when using `arc.strength` in the previous section.

1.6 Using Discrete Bayesian Networks

A BN can be used for inference through either its DAG or the set of local distributions. The process of answering questions using either of these two approaches is known in computer science as *querying*. If we consider a BN as an expert system, we can imagine asking it questions (*i.e.*, querying it) as we would a human expert and getting answers out of it. These answers may take the form of probabilities associated with an event under specific conditions for *conditional probability queries*; they may validate the association between two variables after the influence of other variables is removed for *conditional independence queries*; or they may identify the most likely state of one or more variables for *most likely explanation* queries.

1.6.1   Using the DAG Structure

Using the DAG we saved in dag, we can investigate whether a variable is associated with another, essentially asking a conditional independence query. Both direct and indirect associations between two variables can be read from the DAG by checking whether they are connected in some way. If the variables depend directly on each other, there will be a single arc connecting the nodes corresponding to those two variables. If the dependence is indirect, there will be two or more arcs passing through the nodes that mediate the association. In general, two sets $\mathbf{X}$ and $\mathbf{Y}$ of variables are independent given a third set $\mathbf{Z}$ of variables if there is no set of arcs connecting them that is not *blocked* by the conditioning variables. Conditioning on $\mathbf{Z}$ is equivalent to *fixing* the values of its elements, so that they are known quantities. In other words, the $\mathbf{X}$ and $\mathbf{Y}$ are *separated* by $\mathbf{Z}$, which we denote with $\mathbf{X} \perp\!\!\!\perp_G \mathbf{Y} \mid \mathbf{Z}$. Given that BNs are based on DAGs, we speak of *d-separation* (directed separation): a formal treatment of its definition and properties is provided in Section 6.1. For the moment, we will just say that graphical separation ($\perp\!\!\!\perp_G$) implies probabilistic independence ($\perp\!\!\!\perp_P$) in a BN: if all the paths between $\mathbf{X}$ and $\mathbf{Y}$ are blocked, $\mathbf{X}$ and $\mathbf{Y}$ are (conditionally) independent. The converse is not necessarily true: not every conditional independence relationship is reflected in the graph.

We can investigate whether two nodes in a bn object are d-separated using the dsep function. dsep takes three arguments, x, y and z, corresponding to $\mathbf{X}$, $\mathbf{Y}$ and $\mathbf{Z}$; the first two must be the names of two nodes being tested for d-separation, while the latter is an optional d-separating set. So, for example, we can see from dag that both S and O are associated with R.

```
dsep(dag, x = "S", y = "R")
 [1] FALSE
dsep(dag, x = "O", y = "R")
 [1] FALSE
```

Clearly, S is associated with R because E is influenced by S (S → E) and R is influenced by E (E → R). In fact, the path.exists function shows that there is a path leading from S to R:

```
path.exists(dag, from = "S", to = "R")
 [1] TRUE
```

and, if we condition on E, that path is blocked and S and R become independent.

```
dsep(dag, x = "S", y = "R", z = "E")
 [1] TRUE
```

From Equation (1.1), we can see that indeed the global distribution decomposes cleanly in a part that depends only on S and in a part that depends only on R once E is known:

$$\Pr(\mathsf{S}, \mathsf{R} \mid \mathsf{E}) = \Pr(\mathsf{S} \mid \mathsf{E}) \Pr(\mathsf{R} \mid \mathsf{E}). \tag{1.15}$$

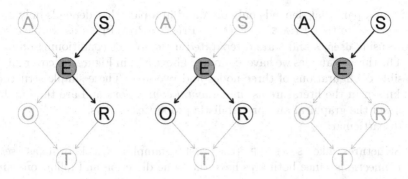

Figure 1.3
Some examples of d-separation covering the three fundamental connections: the *serial connection* (left), the *divergent connection* (centre) and the *convergent connection* (right). Nodes in the conditioning set are highlighted in grey.

The same holds for R and O. They both depend on E, and therefore become independent if we condition on it.

```
dsep(dag, x = "O", y = "R", z = "E")
[1] TRUE
```

Again, from Equation (1.1) we have

$$\Pr(\text{O}, \text{R} \mid \text{E}) = \Pr(\text{O} \mid \text{E}) \Pr(\text{R} \mid \text{E}). \tag{1.16}$$

On the other hand, conditioning on a particular node can also make two other nodes dependent even though they are marginally independent. Consider the following example involving A and S conditional on E.

```
dsep(dag, x = "A", y = "S")
[1] TRUE
dsep(dag, x = "A", y = "S", z = "E")
[1] FALSE
```

From Figure 1.3, we can see that the state of E is influenced by A and S at the same time. Intuitively, if we know what kind of Education one individual has, some combinations of his Age and Sex become more likely than others and, in turn, these two variables become dependent. Equivalently, we can see from Equation (1.1) that E depends on the joint distribution of A and S, as $\Pr(\text{E} \mid \text{A}, \text{S})$; then using Bayes' theorem we have

$$\Pr(\text{E} \mid \text{A}, \text{S}) = \frac{\Pr(\text{A}, \text{S}, \text{E})}{\Pr(\text{A}, \text{S})} = \frac{\Pr(\text{A}, \text{S} \mid \text{E}) \Pr(\text{E})}{\Pr(\text{A}) \Pr(\text{S})} \propto \Pr(\text{A}, \text{S} \mid \text{E}). \tag{1.17}$$

Therefore, when E is known we cannot decompose the joint distribution of A and S in a part that depends only on A and in a part that depends only on S. However, note that $\Pr(A, S) = \Pr(A \mid S) \Pr(S) = \Pr(A) \Pr(S)$: as we have seen above using dsep, A and S are d-separated if we are not conditioning on E.

The three examples we have examined above and in Figure 1.3 cover all the possible configurations of three nodes and two arcs. These simple structures are known in the literature as *fundamental connections* and are the building blocks of the graphical and probabilistic properties of BNs.

In particular:

- Structures like $S \rightarrow E \rightarrow R$ (the first example) are known as *serial connections*, since both arcs have the same direction and follow one after the other.

- Structures like $R \leftarrow E \rightarrow O$ (the second example) are known as *divergent connections*, because the two arcs have divergent directions from a central node.

- Structures like $A \rightarrow E \leftarrow S$ (the third example) are known as *convergent connections*, because the two arcs converge to a central node. When there is no arc linking the two parents (*i.e.*, neither $A \rightarrow S$ nor $A \leftarrow S$) convergent connections are called *v-structures*. As we will see in Chapter 6, their properties are crucial in characterising BNs and learning them from data.

1.6.2   Using the Conditional Probability Tables

In the previous section we have seen how we can answer conditional independence queries using only the information encoded in the DAG. More complex queries, however, require the use of the local distributions. The DAG is still used indirectly, as it determines the composition of the local distributions and reduces the effective dimension of inference problems.

The two most common types of inference are *conditional probability* queries, which investigate the distribution of one or more variables under non-trivial conditioning, and *most likely explanation* queries, which look for the most likely outcome of one or more variables (again under non-trivial conditioning). In both contexts, the variables being conditioned on are the new *evidence* or *findings* which force the probability of an *event* of interest to be re-evaluated. These queries can be answered in two ways, using either *exact* or *approximate inference*; we will describe the theoretical properties of both approaches in more detail in Section 6.6.

1.6.2.1   Exact Inference

Exact inference, which is implemented in package **gRain** (short for "**gRa**phical model **in**ference"), relies on transforming the BN into a specially crafted tree to speed up the computation of conditional probabilities.

```
library(gRain)
```

Such a tree is called a *junction tree* and can be constructed as follows from the bn object we created in the previous section. (The required steps are described in Algorithm 6.4, Section 6.6.2.1, and the junction tree itself is shown in Figure 6.6.)

```
junction <- compile(as.grain(bn))
```

Once the junction tree has been built (by as.grain) and its parameters have been computed (by compile), we can input the evidence into junction using the setEvidence function. The local distributions of the nodes the evidence refers to are then updated, and the changes are propagated through the junction tree. The actual query is performed by the querygrain function, which extracts the distribution of the nodes of interest from junction.

We may be interested, for example, in the attitudes of women towards car and train use compared to the whole survey sample.

```
querygrain(junction, nodes = "T")$T
T
    car  train  other
 0.5618 0.2809 0.1573
jsex <- setEvidence(junction, nodes = "S", states = "F")
querygrain(jsex, nodes = "T")$T
T
    car  train  other
 0.5621 0.2806 0.1573
```

There are no marked differences in the probabilities derived from junction before and after calling setEvidence. The former correspond to $Pr(T)$, the latter to $Pr(T \mid S = F)$. This suggests that women show about the same preferences towards car and train use as the interviewees as a whole.

Another interesting problem is how living in a small city affects car and train use, that is, $Pr(T \mid R = small)$. People working in big cities often live in neighbouring towns and commute to their workplaces because house prices are lower as you move out into the countryside. This, however, forces them to travel mostly by car or train because other means of transport (bicycles, tube, bus lines, etc.) are either unavailable or impractical.

```
jres <- setEvidence(junction, nodes = "R", states = "small")
querygrain(jres, nodes = "T")$T
T
     car   train   other
 0.48389 0.41708 0.09903
```

As shown in Figure 1.4, this reasoning is supported by the BN we saved in the bn object. The probability associated with other drops from 0.1573 to 0.099, while the probability associated with train increases from 0.2808 to 0.4170. Overall, the combined probability of car and train increases from 0.8426 (for the whole survey sample) to 0.9009 (for people living in small

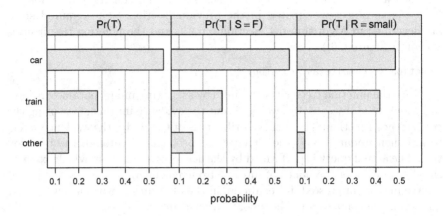

Figure 1.4
Probability distribution of Travel (T) given no evidence (left panel), given evidence that Sex (S) is equal to F (central panel) and given that Residence (R) is equal to small (right panel).

cities). Extending this query to provide the most likely explanation, we conclude that for people living in small cities the car is the preferred means of transport.

Conditional probability queries can also be used to assess conditional independence, as we previously did with graphical separation and the dsep function. Consider again the relationship between S and T, this time conditioning on the evidence that E is equal to high. The joint probability distribution of S and T given E, Pr(S, T | E = high), can be computed using setEvidence and querygrain as follows.

```
jedu <- setEvidence(junction, nodes = "E", states = "high")
SxT.cpt <- querygrain(jedu, nodes = c("S", "T"), type = "joint")
SxT.cpt
   T
S      car   train    other
  M 0.3427 0.1737 0.09623
  F 0.2167 0.1098 0.06087
```

The argument type in querygrain specifies which of the possible distributions involving the nodes is returned. The default value is "marginal", for the marginal distribution of each node.

```
querygrain(jedu, nodes = c("S", "T"), type = "marginal")
$S
S
     M      F
0.6126 0.3874

$T
T
   car  train  other
0.5594 0.2835 0.1571
```

As we have seen above, another possible choice is "joint", for the joint distribution of the nodes. The last valid value is "conditional". In this case querygrain returns the distribution of the first node in nodes conditional on the other nodes in nodes (and, of course, on the evidence we specified with setEvidence).

```
querygrain(jedu, nodes = c("S", "T"), type = "conditional")
    T
S       car  train  other
  M 0.6126 0.6126 0.6126
  F 0.3874 0.3874 0.3874
```

Note how the probabilities in each column sum up to 1, as they are computed conditional on the value that T assumes in that particular column.

Furthermore, we can also see that all the conditional probabilities

$$\Pr(S = M \mid T = t, E = \text{high}), \qquad t \in \{\text{car}, \text{train}, \text{other}\} \qquad (1.18)$$

are identical, regardless of the value of T we are conditioning on, and that the same holds when S is equal to F. In other words,

$$\Pr(S = M \mid T = t, E = \text{high}) = \Pr(S = M \mid E = \text{high}), \qquad (1.19)$$
$$\Pr(S = F \mid T = t, E = \text{high}) = \Pr(S = F \mid E = \text{high}). \qquad (1.20)$$

Hence S is independent from T conditional on E; knowing the Sex of a person is not informative of his preferences if we know his Education. Graphical separation implies that as well since S and T are d-separated by E.

```
dsep(bn, x = "S", y = "T", z = "E")
[1] TRUE
```

Another way of confirming this conditional independence is to use the joint distribution of S and T we stored in SxT.cpt and perform a Pearson's X^2 test for independence. First, we multiply each entry of SxT.cpt by the sample size to convert the conditional probability table into a contingency table.

```
SxT.ct = SxT.cpt * nrow(survey)
```

Each row in survey corresponds to one observation, so nrow(survey) is equal to the sample size. Pearson's X^2 test is implemented in the function chisq.test from package **stats**, which is included in the base R distribution.

```
chisq.test(SxT.ct)

        Pearson's Chi-squared test

data:  SxT.ct
X-squared = 0, df = 2, p-value = 1
```

As expected, we accept the null hypothesis of independence, since the p-value of the test is exactly 1.

1.6.2.2 Approximate Inference

An alternative approach to inference is to use Monte Carlo simulations to randomly generate observations from the BN. In turn, we can use these observations to compute approximate estimates of the conditional probabilities we are interested in. While this approach is computationally expensive, it allows for complex specifications of the evidence and scales better than exact inference to BNs with a large number of nodes.

For discrete BNs, a simple way to implement approximate inference is to use *rejection sampling*. In rejection sampling, we generate random independent observations from the BN. Then we count how many match the evidence we are conditioning on and how many of those observations also match the event whose probability we are computing: the estimated conditional probability is the ratio between the latter and the former.

This approach is implemented in **bnlearn** in the cpquery and cpdist functions. cpquery returns the probability of a specific event given some evidence; so, for example, we can recompute the value of the first cell of the SxT table as follows.

```
cpquery(bn, event = (S == "M") & (T == "car"), evidence = (E == "high"))
[1] 0.3516
```

Note that the estimated conditional probability differs slightly from the exact value computed by querygrain, which is $\Pr(S = M, T = car \mid E = high) = 0.3427$. The quality of the approximation can be improved using the argument n to increase the number of random observations from the default 5000 * nparams(bn) to 10^6.

```
cpquery(bn, event = (S == "M") & (T == "car"),
    evidence = (E == "high"), n = 10^6)
[1] 0.3432
```

The estimated probability is now closer to its true value. However, increasing precision in this way has two drawbacks: answering the query takes much longer, and the precision may still be low if evidence has a low probability.

A better approach is *likelihood weighting*, which will be explained in detail in Section 6.6.2.2. Likelihood weighting generates random observations in such a way that all of them match the evidence, and re-weights them appropriately when computing the conditional probability for the query. It can be accessed from cpquery by setting method = "lw".

```
cpquery(bn, event = (S == "M") & (T == "car"),
  evidence = list(E = "high"), method = "lw")
[1] 0.3422
```

As we can see, cpquery returned a conditional probability (0.3422) that is very close to the exact value (0.3427) without generating 10^6 random observations in the process. Unlike rejection sampling, which is the default for both cpdist and cpquery, evidence for likelihood weighting is provided by a list of values, one for each conditioning variable.

As an example of a more complex query, we can also compute

$$\Pr(\mathsf{S} = \mathsf{M}, \mathsf{T} = \mathsf{car} \mid \{\mathsf{A} = \mathsf{young}, \mathsf{E} = \mathsf{uni}\} \cup \{\mathsf{A} = \mathsf{adult}\}), \qquad (1.21)$$

the probability of a man travelling by car given that his Age is young and his Education is uni or that he is an adult, regardless of his Education.

```
cpquery(bn, event = (S == "M") & (T == "car"),
  evidence = ((A == "young") & (E == "uni")) | (A == "adult"))
[1] 0.3352
```

The implementation of likelihood weighting in cpquery is not flexible enough to compute a query with composite evidence like the above; in that respect it shares the same limitations as the functions in the **gRain** package.

cpdist, which has a syntax similar to cpquery, returns a data frame containing a set of random observations for the variables in nodes that match evidence.

```
SxT <- cpdist(bn, nodes = c("S", "T"), evidence = (E == "high"))
head(SxT)
  S     T
1 M   car
2 M   car
3 M   car
4 M other
5 M   car
6 F other
```

The observations contained in the data frame can then be used for any kind of inference, making this approach extremely versatile. For example, we can produce the probability table of S and T and compare it with that produced by querygrain on page 24. To do that, we first use table to produce a contingency table from the SxT data frame, and then prop.table to transform the counts into probabilities.

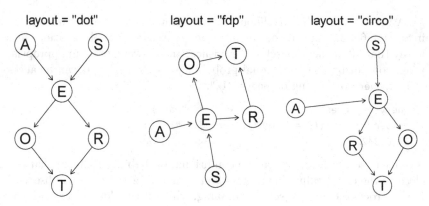

Figure 1.5
Some layouts implemented in **Rgraphviz** and available from graphviz.plot:
dot (the default, on the left), fdp (centre) and circo (right).

```
prop.table(table(SxT))
   T
S      car  train  other
  M 0.3509 0.1833 0.1034
  F 0.1969 0.1042 0.0613
```

Again, we can extend conditional probability queries to produce the most
likely explanation for S and T just by looking for the combination of their
states that has the highest probability. As before, the answer is that among
people whose Education is high, the most common Sex and Travel combination
is male car drivers.

1.7 Plotting Discrete Bayesian Networks

A key strength of BNs, and of graphical models in general, is the possibility of
studying them through their graphical representations. Therefore, the ability
of plotting a BN effectively is a key tool in BN inference.

1.7.1 Plotting DAGs

bnlearn uses the functionality implemented in the **Rgraphviz** package
to plot graph structures, through the graphviz.plot function. If we call
graphviz.plot without any argument other than the graph we want to plot,
we obtain a DAG representation similar to that in Figure 1.1.

```
graphviz.plot(dag)
```

`graphviz.plot` takes care of laying out nodes and arcs so as to minimise their overlap. By default, nodes are positioned so that parents are plotted above their children and that most arcs point downward. This layout is called `dot`. Other layouts can be specified with the `layout` argument; some examples are shown in Figure 1.5.

Highlighting particular nodes and arcs in a DAG, for instance to mark a path or the nodes involved in a particular query, can be achieved either with the `highlight` argument of `graphviz.plot` or using **Rgraphviz** directly. The former is easier to use, while the latter is more versatile.

Consider, for example, the left panel of Figure 1.3. All nodes and arcs with the exception of $S \rightarrow E \rightarrow R$ are plotted in grey, to make the serial connection stand out. The node E, which d-separates S and R, is filled with a grey background to emphasise its role. To create such a plot, we need first to change the colour of all the nodes (including their labels) and the arcs to grey. To this end, we list all the nodes and all the arcs in a list called `hlight`, and we set their `col` and `textCol` to grey.

```
hlight <- list(nodes = nodes(dag), arcs = arcs(dag), col = "grey",
               textCol = "grey")
```

Subsequently, we pass `hlight` to `graphviz.plot` via the `highlight` argument, and save the return value to make further changes to the plot.

```
pp <- graphviz.plot(dag, highlight = hlight, render = FALSE)
```

The `pp` object is an object of class graph, and it can be manipulated with the functions provided by the **graph** and **Rgraphviz** packages. The look of the arcs can be customised as follows using the `edgeRenderInfo` function from **Rgraphviz**.

```
library(Rgraphviz)
edgeRenderInfo(pp) <- list(col = c("S~E" = "black", "E~R" = "black"),
                           lwd = c("S~E" = 3, "E~R" = 3))
```

Attributes being modified (*i.e.*, `col` for the colour and `lwd` for the line width) are specified again as the elements of a list. For each attribute, we specify a list containing the arcs we want to modify and the value to use for each of them. Arcs are identified by labels of the form *parent~child*, *e.g.*, $S \rightarrow E$ is S~E.

Similarly, we can highlight nodes with `nodeRenderInfo`. We set their colour and the colour of the node labels to black and their background to grey.

```
nodeRenderInfo(pp) <-
  list(col = c("S" = "black", "E" = "black", "R" = "black"),
    textCol = c("S" = "black", "E" = "black", "R" = "black"),
    fill = c("E" = "grey"))
```

Once we have made all the desired modifications, we can plot the DAG again with the `renderGraph` function from **Rgraphviz**.

```
renderGraph(pp)
```

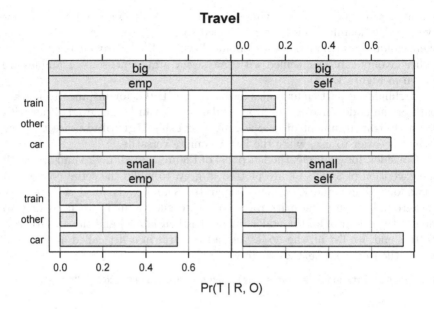

Figure 1.6
Barchart for the probability tables of Travel conditional on Residence and
Occupation.

More complicated plots can be created by repeated calls to `edgeRenderInfo`
and `nodeRenderInfo`. These functions can be used to set several graphical
parameters for each arc or node, and to provide a fine-grained control on
the appearance of the plot. Several (possibly overlapping) groups of nodes
and arcs can be highlighted using different combinations of `lwd` (line width),
`lty` (line type) and `col` (colour); or they can be hidden by setting `col` and
`textCol` to a lighter colour or to `transparent`.

1.7.2 Plotting Conditional Probability Distributions

Plotting the conditional probabilities associated with a conditional probability
table or a query is also useful for diagnostic and exploratory purposes. Such
plots can be difficult to read when a large number of conditioning variables is
involved, but nevertheless they provide useful insights for most synthetic and
real-world data sets.

As far as conditional probability tables are concerned, **bnlearn** provides
functions to plot barcharts (`bn.fit.barchart`) and dot plots (`bn.fit.dotplot`)
from `bn.fit` objects. So, for example, we can produce the plot in Figure 1.6
with

```
bn.fit.barchart(bn.mle$T, main = "Travel",
  xlab = "Pr(T | R, O)", ylab = "")
```

and the corresponding dot plot can be produced by calling bn.fit.dotplot with the same arguments. Each panel in the plot corresponds to one configuration of the levels of the parents of Travel: Occupation and Residence. Therefore, the plot is divided in four panels: {O = self, R = big}, {O = self, R = small}, {O = emp, R = big} and {O = emp, R = small}. The bars in each panel represent the probabilities for car, train and other conditional on the particular configuration of Occupation and Residence associated with the panel.

Both bn.fit.barchart and bn.fit.dotplot use the functionality provided by the **lattice** package, which implements a powerful and versatile set of functions for multivariate data visualisation. As was the case for graphviz.plot and **Rgraphviz**, we can use the **lattice** functions directly to produce complex plots that are beyond the capabilities of **bnlearn**.

Consider, for example, the comparison between the marginal distribution of Travel and the results of the two conditional probability queries shown in Figure 1.4. That plot can be created using the barchart function from **lattice** in two steps. First, we need to create a data frame containing the three probability distributions.

```
Evidence <- factor(c(rep("Unconditional",3), rep("Female", 3),
                  rep("Small City",3)),
            levels = c("Unconditional", "Female", "Small City"))
Travel <- factor(rep(c("car", "train", "other"), 3),
            levels = c("other", "train", "car"))
distr <- data.frame(Evidence = Evidence, Travel = Travel,
          Prob = c(0.5618, 0.2808, 0.15730, 0.5620, 0.2806,
                  0.1573, 0.4838, 0.4170, 0.0990))
```

Each row of distr contains one probability (Prob), the level of Travel it refers to (Travel) and the evidence the query is conditioned on (Evidence).

```
head(distr)
        Evidence Travel  Prob
1 Unconditional    car 0.562
2 Unconditional  train 0.281
3 Unconditional  other 0.157
4        Female    car 0.562
5        Female  train 0.281
6        Female  other 0.157
```

Once the probabilities have been organised in this way, the barchart in Figure 1.4 can be created as follows.

```
library(lattice)
barchart(Travel ~ Prob | Evidence, data = distr,
   layout = c(3, 1), xlab = "probability",
   scales = list(alternating = 1, tck = c(1, 0)),
   strip = strip.custom(factor.levels =
     c(expression(Pr(T)),
       expression(Pr({T} * " | " * {S == F})),
       expression(Pr({T} * " | " * {R == small}})))),
   panel = function(...) {
   panel.barchart(...)
   panel.grid(h = 0, v = -1)
})
```

As can be seen from the first argument of barchart, we are plotting Prob for each level of Travel given the Evidence. For readability, we substitute the label of each panel with an expression describing the probability distribution corresponding to that panel. Furthermore, we lay the panels in a single row (with layout), move all the axes ticks at the bottom of the plot (with scales) and draw a grid over each panel (with the call to panel.grid in the function passed to the panel argument).

Finally, one more plot that may be of general use and that is available in most software packages that implement BNs is a combined DAG-barchart such as that shown in Figure 1.7. Each panel is created with a call to the graphviz.chart function from **bnlearn** as shown below.

```
graphviz.chart(bn, grid = TRUE, main = "Original BN")
graphviz.chart(as.bn.fit(jedu, including.evidence = TRUE), grid = TRUE,
   bar.col = c(A = "black", S = "black", E = "grey", O = "black",
               R = "black", T = "black"),
   strip.bg = c(A = "transparent", S = "transparent", E = "grey",
                O = "transparent", R = "transparent", T = "transparent"),
   main = "BN with Evidence")
```

For each node, graphviz.chart draws a barchart (optionally a dot-plot) of the marginal distribution; and the barcharts are arranged and linked with arcs as the nodes of the DAG of the BN. In the left panel, the BN is shown as we originally defined it at the end of Section 1.3. In the right panel, we converted the jedu BN we produced by introducing the evidence that E = "high" in page 24. We also used some of the optional arguments of graphviz.chart to highlight the node we introduced evidence for. This kind of plot is very useful to see, at a glance, the overall state of the network and how it changes in response to various modelling choices (such as different parameter estimators) or to the introduction of evidence when performing inference.

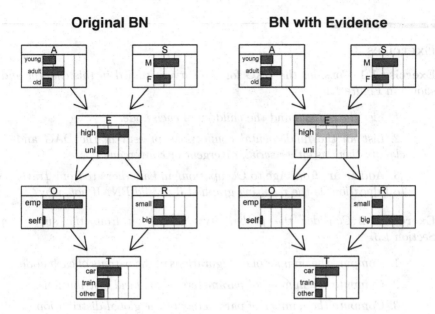

Figure 1.7
Marginal distributions of all the variables in the survey network before (left) and after (right) introducing evidence that E is equal to "high", arranged as a DAG.

1.8 Further Reading

Discrete BNs are the most common type of BN studied in the literature; all the books mentioned in the "Further Reading" sections of this book cover them in detail. Pearl (1988) and Castillo et al. (1997) both explore d-separation in depth. Koller and Friedman (2009, Chapter 17), Korb and Nicholson (2011, Chapter 6) and Neapolitan (2003, Section 7.1) cover parameter learning; Korb and Nicholson (2011, Chapter 9), Koller and Friedman (2009, Chapter 18) and Murphy (2012, Section 16.4) cover structure learning.

Exercises

Exercise 1.1 *Consider the DAG for the survey studied in this chapter and shown in Figure 1.1.*

1. *List the parents and the children of each node.*

2. *List all the fundamental connections present in the DAG and classify them as either serial, divergent or convergent.*

3. *Add an arc from Age to Occupation, and another arc from Travel to Education. Is the resulting graph still a valid BN? If not, why?*

Exercise 1.2 *Consider the probability distribution from the survey in Section 1.3.*

1. *Compute the number of configurations of the parents of each node.*

2. *Compute the number of parameters of the local distributions.*

3. *Compute the number of parameters of the global distribution.*

4. *Add an arc from Education to Travel. Recompute the factorisation into local distributions shown in Equation (1.1). How does the number of parameters of each local distribution change?*

Exercise 1.3 *Consider again the DAG for the survey.*

1. *Create an object of class* bn *for the DAG.*

2. *Use the functions in* **bnlearn** *and the R object created in the previous point to extract the nodes and the arcs of the DAG. Also extract the parents and the children of each node.*

3. *Print the model formula from* bn.

4. *Fit the parameters of the network from the data stored in* survey.txt *using their Bayesian estimators and save the result into an object of class* bn.fit.

5. *Remove the arc from Education to Occupation.*

6. *Fit the parameters of the modified network. Which local distributions change, and how?*

Exercise 1.4 *Re-create the* bn.mle *object used in Section 1.4.*

1. *Compare the distribution of Occupation conditional on Age with the corresponding marginal distribution using* querygrain.

2. *How many random observations are needed for* cpquery *to produce estimates of the parameters of these two distributions with a precision of ±0.01?*

    3. Use the functions in **bnlearn** *to extract the DAG from* `bn.mle`.

    4. Which nodes d-separate Age and Occupation?

Exercise 1.5 *Implement an R function for BN inference via rejection sampling using the description provided in Section 1.4 as a reference.*

Exercise 1.6 *Using the* `dag` *and* `bn` *objects from Sections 1.2 and 1.3:*

    1. Plot the DAG using `graphviz.plot`.

    2. Plot the DAG again, highlighting the nodes and the arcs that are part of one or more v-structures.

    3. Plot the DAG one more time, highlighting the path leading from Age to Occupation.

    4. Plot the conditional probability table of Education.

    5. Compare graphically the distributions of Education for male and female interviewees.

2

The Continuous Case: Gaussian Bayesian Networks

In this chapter we will continue our exploration of BNs, focusing on modelling continuous data under a multivariate Normal (Gaussian) assumption.

2.1 Introductory Example: Crop Analysis

Suppose that we are interested in the analysis of a particular plant, which we will model in a very simplistic way by considering:

- the potential of the plant and of the environment;

- the production of vegetative mass;

- and the harvested grain mass, which is called the *crop*.

To be more precise, we define two synthetic variables to describe the initial status of the plant: its *genetic potential*, which we will denote as G, and the *environmental potential* of the location and the season it is grown in, which we will denote as E. Both G and E are assumed to summarise in a single score all genotypic effects and all environmental effects, respectively; their composite nature justifies the use of continuous variables.

It is well known to farmers and plant breeders that the first step in evaluating a crop is analysing the vegetative organs. Root, stems and leaves grow and accumulate reserves which are later exploited for reproduction. In our model, we will consider a single vegetative variable summarising all the information available on constituted reserves. This variable will be denoted as V, and it will be modelled again as a continuous variable. Clearly, V is directly influenced by the values of G and E. The greater they are, the greater is the possibility of a large vegetative mass V.

As mentioned above, the crop is a function of the vegetative mass. They are related through various quantities, mainly the number of seeds and their mean weight. We will denote them with N and W, respectively. These two variables are not measured at the same time: N is determined at flowering time while W is assessed later in the plant's life. Finally, the crop, denoted by C, depends directly on N and W. All these three variables can again be naturally modelled as continuous variables.

As a result, the behaviour of a plant can be described by

$$\{G, E\} \rightarrow V, \qquad V \rightarrow N, \qquad V \rightarrow W, \qquad \{N, W\} \rightarrow C; \qquad (2.1)$$

and we will use these relationships to model the crop with a BN.

2.2 Graphical Representation

A graphical representation of the relationships in Equation (2.1) is shown in Figure 2.1. The figure displays the six variables used in the BN (the nodes) and the six arcs corresponding to the direct dependencies linking them. Together they form a DAG that is very similar to that presented in Figure 1.1, differing only in the names of the variables. Indeed the DAG does not depend on the nature of the variables under consideration, because the same dependence structure can apply to many different situations.

To work on such a graph, we create an R object describing it. As shown in Section 1.3, we can use the model2network function in **bnlearn** and the model formula representation to express the relationships between the variables.

```
library(bnlearn)
dag.bnlearn <- model2network("[G][E][V|G:E][N|V][W|V][C|N:W]")
dag.bnlearn
```

```
  Random/Generated Bayesian network

  model:
   [E][G][V|E:G][N|V][W|V][C|N:W]
  nodes:                                 6
  arcs:                                  6
    undirected arcs:                     0
    directed arcs:                       6
  average markov blanket size:           2.67
  average neighbourhood size:            2.00
  average branching factor:              1.00

  generation algorithm:                  Empty
```

The structure of the DAG is defined by the string that is the argument of model2network; each node is reported in square brackets along with its parents. As noted in Section 1.3, the order in which the nodes and the parents of each node are given is not important.

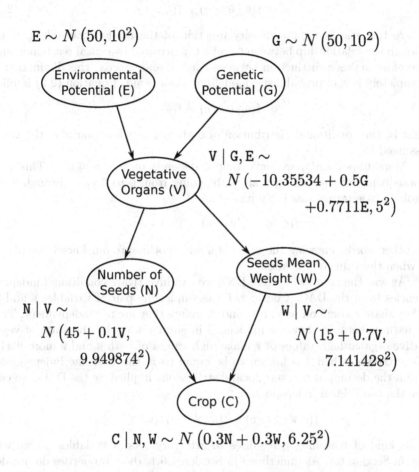

Figure 2.1
DAG representing the crop (C) network, with variables environmental potential (E), genetic potential (G), vegetative organs (V), number of seeds (N) and their mean weight (W). The local probability distributions are shown for each node.

Some key properties of the BN can already be established from the DAG. For instance, in our example we assume that C depends on G because there are paths leading from G to C. From a probabilistic point of view, this means that the conditional distribution of C given $G = g$ is a function of g, or equivalently

$$f(C \mid G = g) \neq f(C), \qquad (2.2)$$

where f(X) stands for the density function of the random variable X. Even though the relationship between C and G is motivated by causal reasoning, and therefore makes a distinction between cause G and effect C, this distinction is completely lost in probabilistic modelling. As a result, Equation (2.2) implies

$$f(G \mid C = c) \neq f(G), \qquad (2.3)$$

that is, the conditional distribution of G given C is a function of c, the value assumed by C.

More importantly, we have that C depends on G through V. This is a consequence of the fact that all paths going from G to C pass through V. In probabilistic terms, this is equivalent to

$$f(C \mid G = g, V = v) = f(C \mid V = v). \qquad (2.4)$$

In other words, knowing the value of G adds nothing to our knowledge about C when the value of V is known.

As was the case for discrete BNs, we can investigate conditional independencies from the DAG. Consider, for example, the pair of variables N and W. They share a common parent, V, and therefore they are not independent. This is natural because they are influenced in similar ways by the state of vegetative organs: high values of V make high values of both N and W more likely. Nevertheless, when V is known to be equal to v, they become independent. From the decomposition into local distributions implied by the DAG, we can see the conditional independence:

$$f(N, W \mid V = v) = f(N \mid V = v) f(W \mid V = v). \qquad (2.5)$$

This kind of reasoning generalises to arbitrary sets of variables, as we will see in Section 6.1. As underlined in Section 1.6.1, these properties do not depend on the nature of the random variables involved, nor on their probability distributions. We can investigate these independencies in a systematic way, whether they are conditional or not, using the dsep function from **bnlearn**.

First, we can find which pairs of variables are marginally independent.

```
crop.nodes <- nodes(dag.bnlearn)
for (n1 in crop.nodes) {
  for (n2 in crop.nodes) {
    if (dsep(dag.bnlearn, n1, n2))
      cat(n1, "and", n2, "are independent.\n")
  }#FOR
}#FOR
```

```
E and G are independent.
G and E are independent.
```

Of all possible pairs, only E and G are found to be independent, and as expected the independence is symmetric. Note that the two nested loops imply that n2 is identical to n1 in some tests and, quite logically, dsep returns FALSE in these cases.

```
dsep(dag.bnlearn, "V", "V")
  [1] FALSE
```

In addition, we can also find which pairs are conditionally independent given V, taking symmetry into account to avoid redundant checks.

```
for (n1 in crop.nodes[crop.nodes != "V"]) {
  for (n2 in crop.nodes[crop.nodes != "V"]) {
    if (n1 < n2) {
      if (dsep(dag.bnlearn, n1, n2, "V"))
        cat(n1, "and", n2, "are independent given V.\n")
    }#THEN
  }#FOR
}#FOR
C and E are independent given V.
C and G are independent given V.
E and N are independent given V.
E and W are independent given V.
G and N are independent given V.
G and W are independent given V.
N and W are independent given V.
```

From the output above we can see that, conditional on the value of V, E and G are no longer independent. On the other hand, nodes belonging to pairs in which one node is *above* V in the graph and the other is *below* V are now conditionally independent. This is due to the d-separating property of V for such pairs of nodes: it descends from the fact that every path connecting two such variables passes through V. Furthermore, we can see that N and W are d-separated as stated in Equation (2.5).

Note that n1 or n2 are chosen so as to never be identical to V. However, dsep accepts such a configuration of nodes and returns TRUE in this case. Since V | V is a degenerate random variable that can assume a single value, it is independent from every random variable.

```
dsep(dag.bnlearn, "E", "V", "V")
  [1] TRUE
```

Many other operations involving the arcs and the nodes of a DAG may be of interest. One possibility is looking for a path going from one subset of nodes to another, which is one way of exploring how the latter depends on the former. For instance, we can use the `path.exists` function to look for the existence of such a path between E and C.

```
path.exists(dag.bnlearn, from = "E", to = "C")
[1] TRUE
```

2.3 Probabilistic Representation

In order to make quantitative statements about the behaviour of the variables in the BN, we need to completely specify their joint probability distribution. This can be an extremely difficult task. In the framework of the multivariate normal distributions considered in this chapter, if we are modelling p variables we must specify p means, p variances and $\frac{1}{2}p\,(p-1)$ correlation coefficients. Furthermore, the correlation coefficients must be such that the resulting correlation matrix is non-negative definite. Fortunately, in the context of BNs we only need to specify the local distribution of each node conditional on the values of its parents, without worrying about the positive definiteness of the correlation matrix of the global distribution. In the case of the DAG shown in Figure 2.1, it means specifying 18 parameters instead of 27.

Suppose for the moment that the local distributions are known and reported in Table 2.1; the fundamental step of estimating the parameter values is postponed to Section 2.4. Of course, the parent sets implied by the conditional distributions are consistent with the DAG, and do not result in any cycle. We can then create a bn object for use with **bnlearn** in the same way we did in Section 1.3. First, we store the parameters of each local distribution in a list, here called `dist.list`.

```
E.dist <- list(coef = c("(Intercept)" = 50), sd = 10)
G.dist <- list(coef = c("(Intercept)" = 50), sd = 10)
V.dist <- list(coef = c("(Intercept)" = -10.35534,
               E = 0.70711, G = 0.5), sd = 5)
N.dist <- list(coef = c("(Intercept)" = 45, V = 0.1), sd = 9.949874)
W.dist <- list(coef = c("(Intercept)" = 15, V = 0.7), sd = 7.141428)
C.dist <- list(coef = c("(Intercept)" = 0, N = 0.3, W = 0.7), sd = 6.25)
dist.list = list(E = E.dist, G = G.dist, V = V.dist,
               N = N.dist, W = W.dist, C = C.dist)
```

Comparing the R definition of each distribution and the corresponding mathematical expression in Table 2.1 elucidates the syntax used in code above.

Subsequently, we can call the `custom.fit` function to create an object of class bn.fit from the graph stored in `dag.bnlearn` and the local distributions in `dist.list`.

```
gbn.bnlearn <- custom.fit(dag.bnlearn, dist = dist.list)
```

$$G \sim N\left(50, 10^2\right)$$
$$E \sim N\left(50, 10^2\right)$$
$$V \mid G = g, E = e \sim N\left(-10.35534 + 0.5g + 0.70711e, 5^2\right)$$
$$N \mid V = v \sim N\left(45 + 0.1v, 9.949874^2\right)$$
$$W \mid V = v \sim N\left(15 + 07v, 7.141428^2\right)$$
$$C \mid N = n, W = w \sim N\left(0.3n + 0.7w, 6.25^2\right)$$

Table 2.1
Probability distributions proposed for the DAG shown in Figure 2.1.

As was the case for multinomial BNs, printing gbn.bnlearn prints all local distributions and their parameters. For brevity, we show only the local distributions of G (a root node) and C (a node with two parents).

```
gbn.bnlearn$G

  Parameters of node G (Gaussian distribution)

Conditional density: G
Coefficients:
(Intercept)
        50
Standard deviation of the residuals: 10
gbn.bnlearn$C

  Parameters of node C (Gaussian distribution)

Conditional density: C | N + W
Coefficients:
(Intercept)            N             W
      0.0            0.3           0.7
Standard deviation of the residuals: 6.25
```

The network we just created rests upon the following assumptions, which characterise linear Gaussian Bayesian networks (GBNs):

- Every node follows a normal distribution.

- Nodes without any parent, known as *root* nodes, are described by the respective marginal (univariate normal) distributions.

- The conditioning effect of the parent nodes is given by an additive linear term in the mean, and does not affect the variance. In other words, each node has a variance that is specific to that node and that does not depend on the values of the parents.

- The local distribution of each node can be equivalently expressed as a Gaussian linear model which includes an intercept and the node's parents as explanatory variables, without any interaction term.

The distributions shown in Table 2.1 would probably not have been an obvious choice for an expert in crop production. Surely, a more complex specification would have been preferred. For instance, using multiplicative terms instead of additive ones in the expectation of $(C \mid N = n, W = w)$ would have been more realistic. The use of linear dependencies is motivated by their good mathematical properties, most importantly their tractability and the availability of closed-form results for many inference procedures. More sophisticated models are certainly possible, and will be presented in Chapters 3 and 5. Nevertheless, it is important to note that for small variations any continuous function can be approximated by an additive function, that is, a first-order Taylor expansion. Furthermore, relatively simple models often perform better than more sophisticated ones when few observations are available.

In the following, we will supplement **bnlearn** with another R package, **rbmn** (short for "réseaux bayésiens multinormaux"), focusing specifically on GBNs. **rbmn** provides a function to convert bn.fit objects such as gbn.bnlearn in its own native format.

```
library(rbmn)
gbn.rbmn <- bnfit2nbn(gbn.bnlearn)
```

It can be shown that the properties assumed above for the local distributions imply that the joint distribution of all nodes (the global distribution) is multivariate normal. Using the decomposition introduced in Section 1.3, we can obtain it as the product of the local distributions:

$$f(G, E, V, N, W, C) = f(G)\, f(E)\, f(V \mid G, E)\, f(N \mid V)\, f(W \mid V)\, f(C \mid N, W). \quad (2.6)$$

The parameters of that multivariate normal distribution can be derived numerically as follows.

```
gema.rbmn <- nbn2gema(gbn.rbmn)
mn.rbmn <- gema2mn(gema.rbmn)
print8mn(mn.rbmn)
    mu s.d.   C.E  C.G   C.V   C.W   C.N   C.C
E 50    10 1.000 0.00 0.707 0.495 0.071 0.368
G 50    10 0.000 1.00 0.500 0.350 0.050 0.260
V 50    10 0.707 0.50 1.000 0.700 0.100 0.520
W 50    10 0.495 0.35 0.700 1.000 0.070 0.721
N 50    10 0.071 0.05 0.100 0.070 1.000 0.349
C 50    10 0.368 0.26 0.520 0.721 0.349 1.000
```

The first column of the `mn.rbmn` matrix is the vector of the marginal expectations (here all equal to 50), the second contains the marginal standard deviations (here all equal to 10), and the remaining columns contain the correlation matrix. The reason behind the choice of these particular values for the parameters in Table 2.1 is now apparent: all marginal distributions have the same mean and variance. This will simplify the interpretation of the relationships between the variables in Section 2.6.

The `mn.rbmn` object is a simple list holding the expectation vector (`mu`) and the covariance matrix (`gamma`) resulting from the correlations and the variances printed above.

```
str(mn.rbmn)
 List of 2
  $ mu   : Named num [1:6] 50 50 50 50 50 ...
  ..- attr(*, "names")= chr [1:6] "E" "G" "V" "W" ...
  $ gamma: num [1:6, 1:6] 100 0 70.71 49.5 7.07 ...
  ..- attr(*, "dimnames")=List of 2
  .. ..$ : chr [1:6] "E" "G" "V" "W" ...
  .. ..$ : chr [1:6] "E" "G" "V" "W" ...
```

Both `mu` and `gamma` have their dimensions names set to the node labels to facilitate the extraction and the manipulation of the parameters of the GBN.

2.4 Estimating the Parameters: Correlation Coefficients

In this section we will tackle the estimation of the parameters of a GBN assuming its structure (*i.e.*, the DAG) is completely known. For this purpose, we generated a sample of 200 observations from the GBN and we saved it in a data frame called `cropdata200`.

```
dim(cropdata200)
 [1] 200   6
round(head(cropdata200), 2)
      C    E    G    N    V    W
 1 48.8 51.5 42.6 54.1 43.0 42.0
 2 48.9 73.4 41.0 60.1 65.3 49.0
 3 67.0 71.1 52.5 51.6 63.2 62.0
 4 37.8 49.3 56.1 49.0 47.8 38.8
 5 55.3 49.3 63.5 54.6 60.6 56.7
 6 56.1 48.7 66.0 44.0 55.5 52.4
```

As we did in Section 1.3 for the discrete case, we can use the `bn.fit` function to produce parameter estimates. **bnlearn** will deduce that the BN is a GBN from the fact that none of the variables in `cropdata200` are factors, and it will use the appropriate estimators for the parameters.

```
crop.fitted <- bn.fit(dag.bnlearn, data = cropdata200)
```

At the time of this writing, `bn.fit` implements only the maximum likelihood estimator. Other estimators can be used in a `bn.fit` object by fitting one model for each node with another R package, and either calling `custom.fit` as we did above or replacing the parameter estimates in an existing `bn.fit` object.

The latter is particularly easy for linear models fitted with the `lm` function; we can just assign the return value of `lm` directly to the corresponding node.

```
crop.fitted$C <- lm(C ~ N + W, data = cropdata200)
```

From the parametric assumptions in Section 2.3, we have that each local distribution can be expressed as a classic Gaussian linear regression model in which the node is the response variable (`C`) and its parents are the explanatory variables (`N`, `W`). The contributions of the parents are purely additive; the model does not contain any interaction term, just the main effect of each parent (`N + W`) and the intercept. The parameter estimates produced by `lm` are the same maximum likelihood estimates we obtained from `bn.fit`. However, it may still be convenient to use `lm` due to its flexibility in dealing with missing values and the possibility of including weights in the model.

Other common forms of regression can be handled in the same way through the **penalized** package (Goeman, 2018). In our example, all the models will have the form $C \sim \mu_C + N\beta_N + W\beta_W$, where μ_C is the intercept and β_N, β_W are the regression coefficients for N and W. Denote each observation for C, N and W with $x_{i,C}$, $x_{i,N}$ and $x_{i,W}$, respectively, where $i = 1, \ldots, n$ and n is the sample size. **penalized** implements *ridge regression*, which estimates β_N, β_W with

$$\{\widehat{\beta}_N^{\text{RIDGE}}, \widehat{\beta}_W^{\text{RIDGE}}\} = \operatorname*{argmin}_{\beta_N, \beta_W} \left\{ \sum_{i=1}^{n} (x_{i,C} - \mu_C - x_{i,N}\beta_N - x_{i,W}\beta_W)^2 + \lambda_2(\beta_N^2 + \beta_W^2) \right\};$$
(2.7)

the *lasso*, with

$$\{\widehat{\beta}_N^{\text{LASSO}}, \widehat{\beta}_W^{\text{LASSO}}\} =$$

$$= \operatorname*{argmin}_{\beta_N, \beta_W} \left\{ \sum_{i=1}^{n} (x_{i,C} - \mu_C - x_{i,N}\beta_N - x_{i,W}\beta_W)^2 + \lambda_1(|\beta_N| + |\beta_W|) \right\};$$
(2.8)

and the *elastic net*, with

$$\{\widehat{\beta}_N^{\text{ENET}}, \widehat{\beta}_W^{\text{ENET}}\} = \operatorname*{argmin}_{\beta_N, \beta_W} \left\{ \sum_{i=1}^{n} (x_{i,C} - \mu_C - x_{i,N}\beta_N - x_{i,W}\beta_W)^2 + \right.$$

$$\left. + \lambda_1(|\beta_N| + |\beta_W|) + \lambda_2(\beta_N^2 + \beta_W^2) \right\},$$
(2.9)

which includes ridge regression and the lasso as special cases when $\lambda_1 = 0$

and $\lambda_2 = 0$, respectively. The parameters λ_1 and λ_2 penalise large values of β_N, β_W in different ways and shrink them towards zero, resulting in smoother estimates and better predictive power. For instance, we can fit ridge regression for C as follows.

```
library(penalized)
crop.fitted$C <- penalized(C ~ N + W, lambda1 = 0, lambda2 = 1.5,
                data = cropdata200)
```

The parameter set for each node of the GBN is stored in one element of the object returned by the bn.fit function. Here is the result for the root node E, that is, a node without any parent: it comprises the expectation and the standard deviation of the node. For reference, the estimated values can be compared with the true values (50, 10) from Table 2.1.

```
crop.fitted$E

  Parameters of node E (Gaussian distribution)

Conditional density: E
Coefficients:
(Intercept)
      50.8
Standard deviation of the residuals: 10.7
```

When a node has one or more parents, the corresponding regression coefficients are also printed.

```
crop.fitted$C

  Parameters of node C (Gaussian distribution)

Conditional density: C | N + W
Coefficients:
(Intercept)           N              W
      2.403       0.273          0.686
Standard deviation of the residuals: 6.31
```

The estimated regression coefficients are close to their true values from Table 2.1, which are $\beta_N = 0.3$ and $\beta_W = 0.7$. The residual standard deviation is also close to its true value, $\sigma_C = 6.25$. The estimated intercept, however, is $\hat{\mu}_C = 2.4026$ and is markedly different from $\mu_C = 0$. We can correct this using lm as above and fitting a model with a null intercept.

```
crop.fitted$C <- lm(C ~ N + W - 1, data = cropdata200)
crop.fitted$C

  Parameters of node C (Gaussian distribution)
```

```
Conditional density: C | N + W
Coefficients:
(Intercept)              N              W
    0.000            0.296          0.711
Standard deviation of the residuals: 6.3
```

Now all the parameters in crop.fitted$C are reasonably close to their true values.

As was the case for discrete BNs, parameter estimates are based only on the subset of the original data frame spanning the considered node and its parents, following the factorisation in Equation (2.6).

```
lmC <- lm(C ~ N + W, data = cropdata200[, c("N", "W", "C")])
coef(lmC)
(Intercept)              N              W
    2.403            0.273          0.686
```

Clearly, the quality of the estimates depends strongly on the sample size.

```
confint(lmC)
               2.5 % 97.5 %
(Intercept) -4.381  9.186
N            0.181  0.366
W            0.589  0.782
```

As we can see, in the case of cropdata200 all the confidence intervals for the parameters of C from Table 2.1 include the corresponding true values.

2.5 Learning the DAG Structure: Tests and Scores

Often expert knowledge on the data is not detailed enough to completely specify the structure of the DAG. In such cases, if sufficient data are available, we can hope that a statistical procedure may help us in determining a small set of conditional dependencies to translate into a sparse BN. In this section, rather than revisiting all the considerations we made at the beginning of Section 1.5, we will concentrate on the tests and scores specific to GBNs.

2.5.1 Conditional Independence Tests

As was the case for discrete BNs, the two classes of criteria used to learn the structure of the DAG are conditional independence tests and network scores. Both are based on statistics derived within the framework of multivariate normality. As far as tests are concerned, the most common is the exact test for *partial correlations*. The usual empirical correlation coefficient, *e.g.*,

$$\rho_{c,w} = \text{COR}(c,w) = \frac{\frac{1}{n}\sum_{i=1}^{n}(x_{i,c} - \bar{x}_c)(x_{i,w} - \bar{x}_w)}{\sqrt{\frac{1}{n}\sum_{i=1}^{n}(x_{i,c} - \bar{x}_c)^2}\sqrt{\frac{1}{n}\sum_{i=1}^{n}(x_{i,w} - \bar{x}_w)^2}} \tag{2.10}$$

can only express marginal linear dependencies between two variables, in this case c and w. However, in GBNs we are often interested in conditional dependencies. Consider the hypothesis that c may be independent from w given N,

$$H_0 : c \perp\!\!\!\perp_P w \mid N \qquad \text{versus} \qquad H_1 : c \not\perp\!\!\!\perp_P w \mid N, \tag{2.11}$$

which is equivalent to setting $\beta_w = 0$ in the regression model we considered in the previous section. The correlation we need to test is the partial correlation between c and w given N, say $\rho_{c,w|N}$, and $c \perp\!\!\!\perp_P w \mid N$ if and only if $\rho_{c,w|N}$ is not significantly different from zero; it can be shown that $\beta_W = 0$ if and only if $\rho_{c,w|N} = 0$, and that holds in general for all regression coefficients in a Gaussian linear model. Unfortunately, there is no closed form expression for partial correlations, but they can be estimated numerically. First, we need to compute the correlation matrix for c, w and N.

```
cormat <- cor(cropdata200[, c("C", "W", "N")])
```

Then we compute the inverse `invcor` of `cormat` with the `cor2pcor` function from package **corpcor** (Schäfer et al., 2017), which works even if the input matrix is not full rank.

```
library(corpcor)
invcor <- cor2pcor(cormat)
dimnames(invcor) <- dimnames(cormat)
invcor
      C      W      N
C 1.000  0.707  0.383
W 0.707  1.000 -0.288
N 0.383 -0.288  1.000
```

We can find the partial correlation $\rho_{c,w|N}$ in `invcor["C", "W"]`. Similarly, $\rho_{c,N|w}$ is in `invcor["C", "N"]` and $\rho_{w,N|c}$ is in `invcor["W", "N"]`. More in general, the (X,Y) element of a partial correlation matrix contains the partial correlation between X and Y given all the other variables.

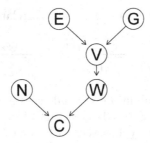

Figure 2.2
The DAG learned from the `cropdata200` data set.

The same estimate for $\rho_{C,W|N}$ is produced by `ci.test` in **bnlearn**, and used to test the hypothesis in Equation (2.11).

```
ci.test("C", "W", "N", test = "cor", data = cropdata200)

        Pearson's Correlation

data:  C ~ W | N
cor = 0.7, df = 197, p-value <2e-16
alternative hypothesis: true value is not equal to 0
```

The distribution for the test under the null hypothesis of independence is a Student's t distribution with $n - 3 = 197$ degrees of freedom for the transformation

$$t(\rho_{C,W|N}) = \rho_{C,W|N}\sqrt{\frac{n-3}{1 - \rho_{C,W|N}^2}}. \tag{2.12}$$

The degrees of freedom are computed by subtracting the number of variables involved in the test (three in this case) from the sample size. If the tested variables are conditionally independent, t is close to zero; large values, either positive or negative, are indicative of the presence and of the direction of the conditional dependence. So, in the case of $\rho_{C,W|N}$, we can say that C has a significant positive correlation with W given N and reject the null hypothesis with an extremely small p-value.

The `cropdata200` data set is not very large, and therefore it is not likely to contain enough information to learn the true structure of the DAG. However, if we perform a naive attempt with one of the algorithms presented in Section 6.5.1.1,

```
pdag1 <- iamb(cropdata200, test = "cor")
```

and compare `pdag1` (shown in Figure 2.2) with Figure 2.1, the result is encouraging: only the arc V → N is missing. All other arcs are present in `pdag1` and their directions are correctly identified.

In order to reduce the number of candidate DAGs and help the structure learning algorithm, **bnlearn** gives the possibility to force the inclusion of some arcs placed in a *whitelist* and to forbid others placed in a *blacklist*. These lists are two-column matrices, similar to those returned by the function arcs.

So if we try the following, we obtain the DAG in Figure 2.1.

```
wl <- matrix(c("V", "N"), ncol = 2)
wl
      [,1] [,2]
[1,] "V"  "N"
pdag2 <- iamb(cropdata200, test = "cor", whitelist = wl)
all.equal(dag.bnlearn, pdag2)
 [1] TRUE
```

Another way to learn a better DAG is to use a bigger sample. Suppose that a sample containing 2×10^4 observations is available in a data frame called cropdata20k.

```
dim(cropdata20k)
 [1] 20000     6
pdag3 <- iamb(cropdata20k, test = "cor")
all.equal(dag.bnlearn, pdag3)
 [1] TRUE
```

Unsurprisingly, we can see from the output above that the DAG has been correctly learned; all arcs are present and in the correct direction. Note that in general some arcs may be undirected regardless of the sample size, because both their directions are equivalent (see Chapter 6 for more details about equivalent DAGs). As a result, a non-causal approach is unable to conclude which one is relevant.

2.5.2 Network Scores

Network scores for GBNs have much in common with the scores for discrete BNs we introduced in Section 1.5.2. For instance, BIC takes the form

$$
\begin{aligned}
\mathrm{BIC} = \log \widehat{f}(\mathsf{E}, \mathsf{G}, \mathsf{V}, \mathsf{N}, \mathsf{W}, \mathsf{C}) - \frac{d}{2} \log n = \\
= \left[\log \widehat{f}(\mathsf{E}) - \frac{d_\mathsf{E}}{2} \log n \right] + \left[\log \widehat{f}(\mathsf{G}) - \frac{d_\mathsf{G}}{2} \log n \right] + \\
+ \left[\log \widehat{f}(\mathsf{V} \mid \mathsf{E}, \mathsf{G}) - \frac{d_\mathsf{V}}{2} \log n \right] + \left[\log \widehat{f}(\mathsf{N} \mid \mathsf{V}) - \frac{d_\mathsf{N}}{2} \log n \right] + \\
+ \left[\log \widehat{f}(\mathsf{W} \mid \mathsf{V}) - \frac{d_\mathsf{W}}{2} \log n \right] + \left[\log \widehat{f}(\mathsf{C} \mid \mathsf{N}, \mathsf{W}) - \frac{d_\mathsf{C}}{2} \log n \right],
\end{aligned} \quad (2.13)
$$

where each local distribution is a normal distribution with the parameters we estimated in Section 2.4. So, for instance,

$$\widehat{f}(C \mid N, W) = N(\widehat{\mu}_C + N\widehat{\beta}_N + W\widehat{\beta}_W, \widehat{\sigma}_C^2), \tag{2.14}$$

where $\widehat{\mu}_C = 0$, $\widehat{\beta}_N = 0.3221$, $\widehat{\beta}_W = 0.6737$ and the residual variance $\widehat{\sigma}_C^2 = 5.8983$. Likewise, the posterior probability score in common use is that arising from a uniform prior over the space of DAGs and of the parameters; it is called the *Bayesian Gaussian equivalent* score (BGe, see Section 6.5 for details). Both scores can be computed by calling the score function from **bnlearn**; the first is obtained with type = "bic-g", the second with type = "bge".

```
score(dag.bnlearn, data = cropdata20k, type = "bic-g")
 [1] -416421.16
score(dag.bnlearn, data = cropdata20k, type = "bge")
 [1] -416494.49
```

BGe accepts two arguments (iss.mu and iss.w) for the imaginary sample sizes associated with the mean and the covariance matrix of the prior distribution, as well as a prior mean vector (nu).

2.6 Using Gaussian Bayesian Networks

As explained in Section 1.6, BNs can be investigated from two different points of view: either focusing on the DAG or using the associated local distributions. DAG properties are independent from the distributional assumptions of the BN. Therefore, we have nothing to add to Section 1.6.1 because the two examples in Figures 1.1 and 2.1 are based on DAGs with the same structure. In this section, we will work only on the local distributions, assuming that the GBN is perfectly known and defined by the gbn.bnlearn object created in Section 2.3. We are interested in the probability of an event or in the distribution of some random variables, usually conditional on the values of other variables. Again, such probabilities can be computed either exactly or approximately, as was the case for discrete BNs.

2.6.1 Exact Inference

Some exact inference procedures are implemented in the **rbmn** package, which we used to compute the global distribution of gbn.bnlearn in Section 2.3. **rbmn** relies on three mathematically equivalent classes: nbn, gema and mn. In an nbn object, the GBN is described by the local distributions, which is natural considering how BNs are defined.

```
print8nbn(gbn.rbmn)
 =====Nodes===[parents]    = Exp. (sd.dev)
 -------------------------------------------------
 ---------E---[-]   = 50  (10)
 ---------G---[-]   = 50  (10)
 ---------V---[E,G]  = -10.355 + 0.707*E + 0.5*G  (5)
 ---------W---[V]   = 15 + 0.7*V  (7.141)
 ---------N---[V]   = 45 + 0.1*V  (9.95)
 ---------C---[N,W]  = 0.3*N + 0.7*W  (6.25)
```

The interpretation of the output is straightforward; the structure of the object can easily be discovered by typing str(gbn.rbmn).

In a gema object, the GBN is described by two generating matrices: a vector of expectations and a matrix to be multiplied by a $N(0,1)$ white noise.

```
print8gema(gema.rbmn)
   mu     E1    E2   E3   E4    E5   E6
 E 50  10.000  0.0 0.0 0.00 0.00 0.00
 G 50   0.000 10.0 0.0 0.00 0.00 0.00
 V 50   7.071  5.0 5.0 0.00 0.00 0.00
 W 50   4.950  3.5 3.5 7.14 0.00 0.00
 N 50   0.707  0.5 0.5 0.00 9.95 0.00
 C 50   3.677  2.6 2.6 5.00 2.98 6.25
```

So, for example, if we consider the row for V in the output above, we can read that

$$V = 50 + 7.071\text{E1} + 5\text{E2} + 5\text{E3}, \qquad (2.15)$$

where $\text{E1}, \text{E2}, \dots, \text{E6}$ are independent and identically distributed $N(0,1)$ Gaussian variables. We already used the mn form in Section 2.3. Note that for gema and mn objects the order of the nodes is assumed to be topological, e.g., parent nodes are listed before their children. Note also that no print functions are associated with these classes and that the print8nbn, print8gema and print8mn must be used instead.

The function condi4joint in **rbmn** can be used to obtain the conditional joint distribution of one or more nodes when the values of some other nodes are fixed. For instance, we can compute the distribution of C when V is fixed to 80 and that of V when C is fixed to 80.

```
print8mn(condi4joint(mn.rbmn, par = "C", pour = "V", x2 = 80))
    mu s.d.
 C 65.6 8.54
print8mn(condi4joint(mn.rbmn, par = "V", pour = "C", x2 = 80))
    mu s.d.
 V 65.6 8.54
```

The results are symmetric because of the normalised distributions we chose. In addition, we can use condi4joint to obtain the conditional distribution of C given an arbitrary value of V by simply not fixing V.

```
unlist(condi4joint(mn.rbmn, par = "C", pour = "V", x2 = NULL))
  mu.C   rho gamma
 24.00  0.52 72.96
```

This means that

$$C \mid V \sim N(24 + 0.52V, 72.9625). \tag{2.16}$$

2.6.2 Approximate Inference

Due to the increasing availability of computational resources, investigating the properties of a given system by simulating it under different conditions and observing its behaviour is becoming increasingly common. When working with probability distributions, "simulation" means the generation of a sample of realisations of random variables. The size of the sample and the approach used to generate it should be chosen in accordance with the magnitude of the probability of the events we want to consider. Clearly, it is much easier to investigate the mean or the median than extreme quantiles in the tails of a distribution: we will see an example of this problem in Section 6.6.2. As shown above, much can be said about a GBN without any simulation. However, for difficult queries simulation is sometimes the only possible approach.

Depending on the query, the simulation can be either direct or constrained. Function rbn implements the former, while cpquery and cpdist provide an easy access to both options. All three functions are in **bnlearn**.

Simulating from a BN, that is, getting a sample of random values from the joint distribution of the nodes, can always be done by sampling from one node at a time using its local distribution, following the order implied by the arcs of the DAG so that we sample from parent nodes before their children. For instance, we can simulate from the nodes following the order of Table 2.1, using the values simulated in previous steps for conditioning. In such a simulation, the values generated for each node are a sample from its marginal distribution. This is true regardless of the nature and the distributions of the nodes. In addition, the same global simulation can be used for a pair of nodes, such as (V, N), as well as for any other subset of nodes. For instance, we can generate nobs = 4 observations from (V, N) using our crop GBN as follows.

```
nobs <- 4
VG <- rnorm(nobs, mean = 50, sd = 10)
VE <- rnorm(nobs, mean = 50, sd = 10)
VV <- rnorm(nobs, mean = -10.355 + 0.5 * VG + 0.707 * VE, sd = 5)
VN <- rnorm(nobs, mean = 45 + 0.1 * VV, sd = 9.95)
cbind(VV, VN)
        VV   VN
[1,] 43.8 45.6
[2,] 48.0 62.1
[3,] 50.1 52.0
[4,] 54.3 40.9
```

Of course, in the case of GBNs it is quicker and easier to use **bnlearn**:

```
sim <- rbn(gbn.bnlearn, n = nobs)
sim[, c("V", "N")]
      V    N
1 44.2 41.7
2 50.5 50.7
3 49.0 59.4
4 29.4 46.8
```

In fact the two data sets we used to introduce the statistical procedures in Sections 2.4 and 2.5 were produced in this way, with the R code:

```
set.seed(4567)
cropdata200 <- rbn(gbn.bnlearn, n = 200)
set.seed(1234)
cropdata20k <- rbn(gbn.bnlearn, n = 20000)
```

For the moment, we have not introduced any restriction on the simulation, even though it is common to do so in practice. For instance, we may discuss with an agronomist about our choices in modelling the vegetative mass V. As a result, we become interested in simulating V in such extreme scenarios as $V \mid G = 10, E = 90$ to see what happens for particularly bad genotypes grown in very favourable environments. In this case, by construction, we know the conditional distribution we need to answer this question: it is given in the third line of Table 2.1. It is also possible to get the distribution of a pair of conditioned variables, *e.g.*, $N, W \mid V = 35$. This can be achieved by transforming (or building) the corresponding GBN. In addition, when the conditioning is not given directly but we can write it in closed form as in Equation (2.16), it is possible to simulate directly from the conditional distribution.

But this is not always the case. For instance, suppose that we are interested in the conditional distribution $N, W \mid C > 80$ associated with the following question: *what are the values of N and W associated with a very good crop?* To answer that we need to condition on an interval, not on a single value.

A naive but correct approach is: *just make a simulation with a high number of draws, and retain only those satisfying the condition,* here $C > 80$. However, that it is not feasible when the probability of generating observations satisfying the condition is very small. Anyway, we can try that way with cpquery and cpdist.

```
head(cpdist(gbn.bnlearn, nodes = c("C", "N", "W"), evidence = (C > 80)))
     C    N    W
1 83.9 60.3 77.7
2 81.6 65.3 81.7
3 83.6 64.8 74.4
4 80.7 43.8 70.1
5 80.0 73.5 59.3
6 83.4 67.8 65.0
```

Such an approach is clearly not possible when we are conditioning on a single value for one or more variables, as in the case of V | G = 10, E = 90. In continuous distributions, a single value always has probability 0; only intervals may have a non-zero probability. As a result, we would discard all the samples we generate! We need a more advanced simulation approach to handle this case. A simple one is likelihood weighting from Section 1.6.2.2 (detailed in Section 6.6.2), which can be accessed from cpdist by setting method = "lw".

```
head(cpdist(gbn.bnlearn, nodes = "V",
        evidence = list(G = 10, E = 90), method = "lw"), n = 5)
    V
1 53.0
2 56.2
3 77.3
4 57.3
5 61.6
```

As we can see from the code above, the evidence for this method is provided by a list of values, one for each conditioning variable. Similarly, the probability of a specific event can be computed using likelihood weighting via cpquery. So, for example, we may be interested in the probability of having a vegetative mass above 70 in the conditions specified by G and E.

```
cpquery(gbn.bnlearn, event = (V > 70),
    evidence = list(G = 10, E = 90), method = "lw")
[1] 0.00978
```

The probability we obtain is very low despite the favourable environments, as expected from such bad genotypes.

2.7 Plotting Gaussian Bayesian Networks

2.7.1 Plotting DAGs

Plots displaying DAGs can be easily produced using several different R packages such as **bnlearn**, **Rgraphviz**, **igraph** and others. In this chapter, we will concentrate on **igraph** since **bnlearn** and **Rgraphviz** have already been presented in Chapter 1. Nevertheless we think that the most convenient way for **bnlearn** users is the function graphviz.plot which provides an interface with the R package **Rgraphviz** as shown in Section 1.7.1.

First, we introduce how to define a DAG via its arc set with **igraph**.

```
library(igraph)
igraph.options(print.full = TRUE)
dag0.igraph <- graph.formula(G-+V, E-+V, V-+N, V-+W, N-+C, W-+C)
dag0.igraph
```

```
IGRAPH af499ba DN-- 6 6 --
+ attr: name (v/c)
+ edges from af499ba (vertex names):
[1] G->V V->N V->W E->V N->C W->C
```

The arguments provided to the graph.formula function identify the nodes at the tail and at the head of each arc in the graph. For instance, E-+V indicates that there is an arc going from node E to node V. The "-" sign means that E is at the tail of the arc, while the "+" means that V is at the head. Therefore, E-+V and V+-E identify the same arc.

Starting from the objects we generated with **bnlearn**, we first need to convert a bn or bn.fit object into an **igraph** graph object.

```
dag.igraph <- as.igraph(dag.bnlearn)
```

Even though **igraph** implements a large number of functions for handling graphs (not only DAGs), only a very small part will be shown here. We refer the interested reader to the extensive documentation included in the package. Among these functions, V and E display the *vertices* and the *edges* of a graph, which are synonymous with nodes and arcs when working with DAGs.

```
V(dag.igraph)
+ 6/6 vertices, named, from 463dd3d:
[1] C E G N V W
E(dag.igraph)
+ 6/6 edges from 463dd3d (vertex names):
[1] G->V E->V V->N V->W N->C W->C
```

We can now plot our DAG in different ways by applying some of the layout algorithms available in **igraph**. Here is the R code which produces Figure 2.3.

```
par(mfrow = c(1, 3), mar = rep(3, 4), cex.main = 2)
plot(dag.igraph, main = "\n1: defaults")
ly <- matrix(c(2, 3, 1, 1, 2, 3, 1, 4, 4, 2, 3, 2), 6)
plot(dag.igraph, layout = ly, main = "\n2: positioning")
vcol <- c("black", "darkgrey", "darkgrey", rep(NA, 3))
lcol <- c(rep("white", 3), rep(NA, 3))
par(mar = rep(0, 4), lwd = 1.5)
plot(dag.igraph, layout = ly, frame = TRUE, main = "\n3: final",
     vertex.color = vcol, vertex.label.color = lcol,
     vertex.label.cex = 3, vertex.size = 50,
     edge.arrow.size = 0.8, edge.color = "black")
```

When drawing DAGs, it is often useful to manually position the nodes to highlight the structure, the logic and the purpose of the BN. Packages such as **graph** and **Rgraphviz** provide several automatic graph drawing algorithms, but for small graphs the best solution is for the user to place each node. With **igraph**, a convenient way to plot graphs is to call tkplot first, hand-tune

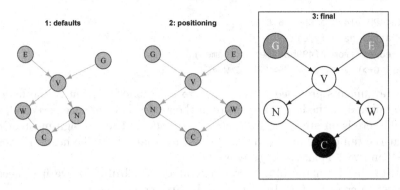

Figure 2.3
Three plots of the crop DAG obtained by specifying more and more arguments
with the **igraph** package.

the placement of the nodes, query the coordinates with the `tkplot.getcoords`
function and use them to `plot` the graph.

We illustrate how to customise a plot step by step in Figure 2.3. In the
first panel, we plot the DAG with the default arguments. In the second panel,
we fix the position of the nodes with a two-columns matrix. Finally, we add
some formatting in the third panel. The code above shows how arguments can
be specified either globally, like the colours of the arcs, or individually, like
the colour of the nodes (`NA` means *no colour*).

2.7.2 Plotting Conditional Probability Distributions

bnlearn does not provide any function to plot local probability distributions
in GBNs, unlike the `bn.fit.barchart` function available for discrete BNs. The
reason is that while the parents of a node in a discrete BN have a finite number
of configurations, which are trivial to enumerate and plot, the parents of a
node in a GBN are defined over $\mathbb{R}$ and the corresponding local distribution
is very difficult to plot effectively. However, some of common diagnostic plots
for linear regression models are available:

- `bn.fit.qqplot`: a quantile-quantile plot of the residuals;

- `bn.fit.histogram`: a histogram of the residuals, with their theoretical normal
  density superimposed;

- `bn.fit.xyplot`: a plot of the residuals against the fitted values.

All these functions are based on the **lattice** functions of the same name and
can be applied to a `bn.fit` object, thus producing a plot with one panel for
each node.

```
gbn.fit <- bn.fit(dag.bnlearn, cropdata20k)
bn.fit.qqplot(gbn.fit)
```

They can plot a single node in a bn.fit object as well.

```
bn.fit.qqplot(gbn.fit$V)
```

It is important to note that such plots require the residuals and the fitted values to be stored in the bn.fit object. Therefore, GBNs created with custom.fit will produce an error unless both quantities have been provided by the user.

```
bn.fit.qqplot(gbn.bnlearn)
 Error in lattice.gaussian.backend(fitted = fitted, type = "qqplot",
 xlab = xlab, : no residuals present in the bn.fit object.
```

As far as other conditional distributions are concerned, it is not difficult to produce some interesting plots. Here we will give an example with **rbmn**. Suppose that we are interested in how C changes in response to variations in E and V, that is, in C | E, V. In fact, due to the good properties of GBNs, we can derive the closed form of the associated distribution using:

```
C.EV <- condi4joint(mn.rbmn, par = "C", pour = c("E", "V"), x2 = NULL)
C.EV$rho
   E    V
 C 0 0.52
```

The zero regression coefficient obtained for E when V is introduced underscores how no additional information is added by E once V is already known. This is because V d-separates E and C:

```
dsep(gbn.bnlearn, "E", "C", "V")
 [1] TRUE
```

But imagine that we are in a more complicated case (see Chapter 3 for some examples) in which such an approach is not possible. Our aim is to produce a plot providing insight on the distribution of C when both E and V vary. Since creating a three-dimensional plot is difficult and error-prone, we will replace the third dimension with the size of the points representing each simulated observation. This can be done with the R commands below.

```
set.seed(5678)
cropdata3 <- cpdist(gbn.bnlearn, nodes = c("E", "V", "C"),
                    evidence = TRUE, n = 1000)
plot(cropdata3$V, cropdata3$C, type = "n",
     main = "C | V, E; E is the point size")
cexlim <- c(0.1, 2.4)
cexE <- cexlim[1] + diff(cexlim) / diff(range(cropdata3$E)) *
                    (cropdata3$E - min(cropdata3$E))
points(cropdata3$V, cropdata3$C, cex = cexE)
```

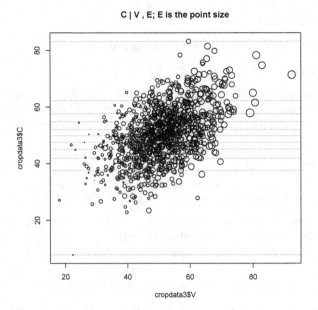

Figure 2.4
Simulated distribution of C given E and V. Horizontal lines correspond to the deciles of C (*e.g.*, the 0%, 10%, 20%, ..., 100% quantiles), so there is the same number of points in each horizontal slice.

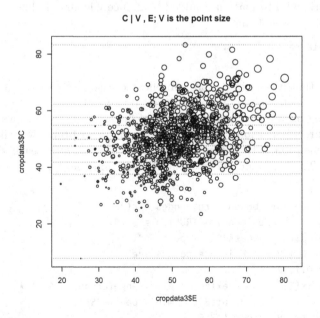

Figure 2.5
Simulated distribution of C given E and V, in the same format as Figure 2.4.

```
cqa <- quantile(cropdata3$C, seq(0, 1, 0.1))
abline(h = cqa, lty = 3)
```

The result is shown in Figure 2.4. We can see a strong relationship between V and C, the variables in the x and y axes. No additional effect from E to V is apparent: for any given level of C, the variation of both variables is about the same. Changing their roles (Figure 2.5) highlights the additional effect of V with respect to E.

2.8 More Properties

Much more could be said about the properties of GBNs. Those that arise from graphical models theory can be found in Chapter 6. General properties of multivariate normal distributions, and thus of the global and local distributions in a GBN, can be found in classic multivariate statistics books such as Mardia et al. (1979) and Anderson (2003).

In this last section, we would like to recall some of the most remarkable.

1. The precision matrix is the inverse of the covariance matrix of the global distribution. When the DAG is moralised (see Section 6.4), the absence of an arc between two nodes implies a zero entry in the precision matrix and vice versa. This is because the (i, j) entry of the precision matrix is the partial correlation between the ith and the jth variables given the rest, that is,

$$\rho_{X_i, X_j | \mathbf{X} \setminus \{X_i, X_j\}} = \mathrm{COR}(X_i, X_j \mid \mathbf{X} \setminus \{X_i, X_j\}),$$

and $\rho_{X_i, X_j | \mathbf{X} \setminus \{X_i, X_j\}}$ is equal to zero if and only if the regression coefficient of X_j against X_i is zero as well.

2. If some nodes are just linear combinations of their parents, their conditional standard deviation is zero and the global distribution is degenerate because its covariance matrix is not full rank. In that case, such nodes must be removed before computing the precision matrix; their children can be made to depend on the removed nodes' parents to preserve the original dependencies in a transitive way. In fact, these deterministic nodes do not have any influence on the GBN's behaviour and they can be safely disregarded.

3. Furthermore, the global distribution can be singular or numerically ill-behaved if the GBN is learned from a sample whose size is smaller than the number of parameters ($n \ll p$) or not large enough ($n \approx p$). Since all the optimality properties of the covariance matrix are asymptotic, they hold only approximately even for very large ($n \gg p$) samples. The need for $n \gg p$ can be obviated by the use of penalised, shrinkage or Bayesian estimation techniques that supplement the lack of information in the data and enforce sparsity and regularity in the GBN. More on this topic will be covered in Chapter 6.

2.9   Further Reading

For a broader overview of GBNs, basic definitions and properties are in Koller and Friedman (2009, Chapter 7). Constraint-based structure learning is explored in Korb and Nicholson (2011, Chapter 8), parameter learning in Neapolitan (2003, Section 7.2) and inference in Neapolitan (2003, Section 4.1).

Exercises

Exercise 2.1 *Prove that Equation (2.2) implies Equation (2.3).*

Exercise 2.2 *Within the context of the DAG shown in Figure 2.1, prove that Equation (2.5) is true using Equation (2.6).*

Exercise 2.3 *Compute the marginal variance of the two nodes with two parents from the local distributions proposed in Table 2.1. Why is it much more complicated for C than for V?*

Exercise 2.4 *Write an R script using only the* rnorm *and* cbind *functions to create a* 100×6 *matrix of 100 observations simulated from the BN defined in Table 2.1. Compare the result with those produced by a call to* cpdist *function.*

Exercise 2.5 *Imagine two ways other than changing the size of the points (as in Section 2.7.2) to introduce a third variable in the plot.*

Exercise 2.6 *Can GBNs be extended to log-normal distributions? If so how, if not, why?*

Exercise 2.7 *How can we generalise GBNs as defined in Section 2.3 in order to make each node's variance depend on the node's parents?*

Exercise 2.8 *From the first three lines of Table 2.1, prove that the joint distribution of E, G and V is trivariate normal.*

3

The Mixed Case: Conditional Gaussian Bayesian Networks

In Chapters 1 and 2 we considered BNs with either discrete or continuous variables. Moreover, in each BN all variables followed probability distributions belonging to the same family: multinomial or normal.

In this chapter, we will cover how these two families can be combined to create a *conditional Gaussian* BN (CGBN). A CGBN is a "mixture of normals" model in which continuous nodes can have both continuous and discrete parents, while discrete nodes can only have discrete parents. We see this as an initial step towards the more complex BNs presented in Chapters 4 and 5, which provide even greater flexibility.

3.1 Introductory Example: Healthcare Costs

The cost of healthcare is a recurring theme in most countries' public discourse due to the combination of an ageing population and the availability of more advanced (read, expensive) treatments. Inspired by Chao et al. (2017) and by the more recent Wang et al. (2019), we will try to build a simple BN to model an individual's yearly medical expenditure. We will use the UK National Health Service (NHS) as an inspiration: it is free at the point of use for most things and it is financed through taxes (the National Insurance and general taxation).

For the sake of the example, we consider only seven variables:

- **Age** (A, discrete): the age, recorded as *young* (young), *adult* (adult) or *old* (old) as in the train survey in Chapter 1.

- **Pre-existing conditions** (C, discrete): whether the individual has *no pre-existing conditions* (none), *mild* (mild) or *severe pre-existing conditions* (severe).

- **Outpatient expenditure** (O, continuous): the cost of the individual's outpatient hospital visits, like specialist consultations.

- **Inpatient expenditure** (I, continuous): the cost of the individual's inpatient hospital visits, which include hospital admissions.

- **Any hospital say** (H, discrete): whether the individual spent any days at the hospital, recorded as none or any.

- **Days of hospital stay** (D, continuous): how many days the individual spent at the hospital.

- **Taxes** (T, continuous): the amount of money the individual should pay in taxes to cover his medical expenses.

Pre-existing conditions become more likely with age: older people are inherently more fragile than younger people. For the same reason, we expect older people to have both higher outpatient and inpatient expenditures; and to be admitted for longer periods of time during hospital stays. Both expenditures over the course of one year should be predictive of the overall expenditures for the following year, and therefore should be used to determine the amount of tax paid. As a result, we will structure the BN as shown in Figure 3.1 using the following relationships:

$$A \to O, \qquad A \to C, \qquad A \to H, \qquad \{A, H\} \to D, \qquad \{C, D\} \to I, \qquad \{O, I\} \to T.$$

Note that we explicitly do not include the arcs $A \to T$ and $C \to T$, since it is illegal to discriminate individuals by either their age or pre-existing conditions in most settings (including health insurance).

3.2   Graphical and Probabilistic Representation

The DAG encoding the relationship described above can easily be created using model2network, similarly to what have done in Chapters 1 and 2.

```
dag <- model2network("[A][C|A][H|A][D|A:H][I|C:D][O|A][T|O:I]")
```

As for the local distributions of A, C, H, D, I, O and T, we choose to model discrete variables with multinomial random variables as in Chapter 1; and to model continuous variables with Gaussian distributions, which we parameterise as regression models following Chapter 2. In order to find realistic values for the parameters (conditional probabilities for discrete variables, regression coefficients and standard errors for continuous variables) we will look into publicly available information from NHS England.[1]

First of all, we can assign probabilities to the three age brackets of A using the 2011 UK census data.

```
A.lv <- c("young", "adult", "old")
A.prob <- array(c(0.35, 0.45, 0.20), dim = 3, dimnames = list(A = A.lv))
```

---

[1]Of which there is plenty at https://www.england.nhs.uk/ourwork/tsd/data-info/open-data

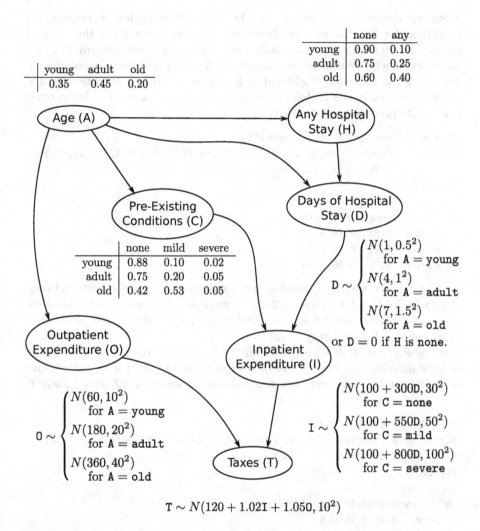

Figure 3.1
DAG representing the healthcare costs network, with the local distributions of each node.

```
A.prob
A
young adult   old
 0.35  0.45  0.20
```

Then we define the prevalence of the different severities of pre-existing conditions for C in different age brackets. NHS England tells us that 58% of people over 60 have long-term conditions or chronic diseases compared to 14% under 40, which gives the proportion of people with mild or severe conditions. To reflect the increased likelihood of having some sort of condition as age increases, we make the probability of having a mild or severe condition twice that in the preceding age bracket.

```
C.lv <- c("none", "mild", "severe")
C.prob <- array(c(0.88, 0.10, 0.02, 0.75, 0.20, 0.05, 0.42, 0.53, 0.05),
           dim = c(3, 3), dimnames = list(C = C.lv, A = A.lv))
C.prob
         A
C        young adult  old
  none     0.88  0.75 0.42
  mild     0.10  0.20 0.53
  severe   0.02  0.05 0.05
```

The average length of hospital stay for admitted patients varies between 5 days in 2014 and 4.5 days in 2019. Considering that younger people have fewer and shorter hospital stays than older people, we choose the probability of having any hospital stay at all (that is, H is "none") to be very low when A == "young" and to increase significantly with age. We then choose the values of D for different age brackets to be always equal to zero if H == "none", and to gradually increase with age while averaging between 4.5 and 5 days if H == "any".

```
H.lv <- c("none", "any")
H.prob <- array(c(0.90, 0.10, 0.75, 0.25, 0.60, 0.40), dim = c(2, 3),
           dimnames = list(H = H.lv, A = A.lv))
H.prob
       A
H       young adult old
  none   0.9  0.75 0.6
  any    0.1  0.25 0.4
D.coef <- list(coef = array(c(0, 0, 0, 1, 4, 7), dim = c(1, 6),
                    dimnames = list("(Intercept)", NULL)),
           sd = c(0, 0, 0, 0.5, 1, 1.5))
D.coef
$coef
            [,1] [,2] [,3] [,4] [,5] [,6]
(Intercept)   0    0    0    1    4    7
```

```
$sd
[1] 0.0 0.0 0.0 0.5 1.0 1.5
```

The elements of D.coef$coef correspond to the configurations of H × A; those that correspond to H == "none" are set to zero as discussed above. Naming the elements is not required because custom.fit will take care of it, removing any names we might have assigned in the process.

In mathematical notation, D is distributed as a degenerate $N(0,0)$ if H == "none"; hence it is always equal to zero., If H == "any", the local distribution of D can be written equivalently as a *mixture of normal distributions*:

$$D \sim \begin{cases} N(1, 0.5^2) & \text{for A} = \text{young} \\ N(4, 1^2) & \text{for A} = \text{adult} \\ N(7, 1.5^2) & \text{for A} = \text{old} \end{cases} \tag{3.1}$$

or as a *mixture of linear regressions*:

$$\begin{cases} D = 1 + \varepsilon_{\text{young}} \\ D = 4 + \varepsilon_{\text{adult}} \\ D = 7 + \varepsilon_{\text{old}} \end{cases} \quad \text{with} \quad \begin{cases} \varepsilon_{\text{young}} \sim N(0, 0.5^2) \\ \varepsilon_{\text{adult}} \sim N(0, 1^2) \\ \varepsilon_{\text{old}} \sim N(0, 1.5^2) \end{cases} \tag{3.2}$$

In other words, if a continuous variable has a discrete variable as a parent the local distribution of that variable is a set of linear regression models (or equivalently, normal distributions) with one regression for each value of the discrete parent. If a continuous variable (like D) has more than one discrete parent (A and H), the set will contain a distinct regression for each configuration of their values.

But what if a continuous variable also has continuous parents? They are included in all the regressions in the mixture as additive effects, with different coefficients in each regression. Hence in the case of I, the inpatient expenditures, the regression coefficients are organised in a matrix with one column for each configuration of the discrete parents (*i.e.*, C) and one row for each coefficient in the linear regressions.

```
I.coef <- list(coef = array(c(1, 3, 1, 5.5, 1, 8) * 100, dim = c(2, 3),
                     dimnames = list(c("(Intercept)", "D"), NULL)),
            sd = c(30, 50, 100))
I.coef
$coef
            [,1] [,2] [,3]
(Intercept)  100  100  100
D            300  550  800

$sd
[1]  30  50 100
```

We choose the intercept to encode the fixed admin costs associated with a hospital stay (say £100) with the intercept, and the per-day costs of care with the regression coefficient for D. NHS data tell us that the average per-day cost of a hospital stay is £400: we assume that patients with no pre-existing conditions have lower costs than patients with severe pre-existing conditions, which presumably require more care as well as more expensive medications and equipment. We might have chosen different intercepts for different values of A as well, implying different fixed costs for patients with pre-existing conditions. We choose not to for simplicity. In mathematical terms, we have

$$\begin{cases} \text{I} = 100 + 300\text{D} + \varepsilon_{none} \\ \text{I} = 100 + 500\text{D} + \varepsilon_{mild} \\ \text{I} = 100 + 800\text{D} + \varepsilon_{severe} \end{cases} \quad \text{with} \quad \begin{cases} \varepsilon_{none} \sim N(0, 30^2) \\ \varepsilon_{mild} \sim N(0, 50^2) \\ \varepsilon_{severe} \sim N(0, 100^2) \end{cases} \quad (3.3)$$

In comparison, outpatient expenditures are smaller; and their distribution is simpler since O only depends on the age A.

```
O.coef <- list(coef = array(c(60, 180, 360) , dim = c(1, 3),
                    dimnames = list("(Intercept)", NULL)),
            sd = c(10, 20, 40))
```

Finally, we can compute the amount of tax T as an additive function of some fixed costs (the intercept) and of the inpatient and outpatient expenditures:

```
T.coef <- list(coef = c("(Intercept)" = 120, I = 1.02, O = 1.05),
            sd = 10)
```

which can formally be written as $\text{T} = 120 + 1.02\text{I} + 1.05\text{O} + \varepsilon_T \sim N(0, 10^2)$. Clearly, the regression coefficients associated with I and O must be larger than 1, since we need to collect enough taxes to pay for the expected amount of healthcare expenditures. In an ideal world, excess taxes could be used for unforeseen emergencies or reinvested back into the healthcare system.

Having defined the DAG and the local distributions for all the variables in the BN, we can assemble the bn.fit object encoding the BN using custom.fit.

```
dists <- list(A = A.prob, C = C.prob, H = H.prob, D = D.coef,
            I = I.coef, O = O.coef, T = T.coef)
healthcare <- custom.fit(dag, dists)
```

For brevity, we will only print the local distribution of I to show how a mixture of linear regression is displayed.

```
healthcare$I

  Parameters of node I (conditional Gaussian distribution)

  Conditional density: I | C + D
  Coefficients:
```

```
                 0    1    2
(Intercept)    100  100  100
D              300  550  800
Standard deviation of the residuals:
   0    1    2
  30   50  100
Discrete parents' configurations:
         C
0    none
1    mild
2  severe
```

Note how the regression coefficients are organised in a matrix much like that we created above, but with a better labelling of the columns. The configurations of the discrete parents are numbered (starting from zero) and are listed at the end, because concatenating the levels of multiple discrete parents can result in pretty long labels.

The BN we just created is called a *conditional Gaussian BN* (CGBN); it is also known as a *conditional linear Gaussian BN*. CGBNs make the following assumptions:

- Discrete nodes follow a multinomial distribution.

- Continuous nodes that do not have any discrete node among their parents follow a normal distribution.

- Continuous nodes that have one or more discrete nodes among their parents follow a mixture of normal distributions with one component for each combination of the values of those discrete parents.

- Each normal distribution in each mixture has a separate mean and variance, that is, each component of the mixture has an independent set of parameters.

- Continuous nodes are allowed to have both continuous nodes and discrete nodes as parents, but discrete nodes can only have other discrete nodes as parents.

3.3 Estimating the Parameters: Mixtures of Regressions

Estimating the parameters of a CGBN when its DAG is known is an extension of what we have seen in Sections 1.4 and 2.4. As before, we first read a data set from a file; and we take care of telling read.table which columns are numeric and which are factors using the colClasses argument.

```
costs <- read.table("healthcare.txt", header = TRUE,
        colClasses = c("factor", "factor", "numeric", "factor",
                       "numeric", "numeric", "numeric"))
```

We can then use costs and the dag we created in the previous section to estimate the parameters of the local distributions using maximum likelihood.

```
fitted <- bn.fit(dag, data = costs)
```

The form of the maximum likelihood estimators depends on the distribution we assume for each variable and on whether that variable has any discrete parents. In the case of discrete variables, the conditional probabilities of a node (say, H) given its parents are estimated using empirical frequencies as in Section 1.4.

```
cpt.H <- prop.table(table(costs[, c("H", "A")]), margin = 2)
all.equal(cpt.H, coef(fitted$H))
  [1] TRUE
```

As for continuous nodes with no discrete parent, but possibly some continuous parents, the parameters of the local distributions are the regression coefficients associated with the parents and the standard deviation of the residuals. As in Section 2.4, they are identical to those estimated by lm.

```
params.T <- lm(T ~ I + 0, data = costs)
all.equal(coef(fitted$T), coef(params.T))
  [1] TRUE
all.equal(sigma(fitted$T), sigma(params.T))
  [1] TRUE
```

Finally, continuous nodes with at least one discrete parent are modelled with a mixture of linear regressions. Take, for example, the inpatient expenditure I: it has one discrete parent (C) and one continuous parent (D). We can compute the maximum likelihood parameter estimates for such a mixture by splitting the data into the subsets corresponding to the three values of C ("none", "mild", "severe") and fitting a separate linear model for each subset using lm. The regression coefficients then are:

```
models.I <- list(lm(I ~ D, data = costs[costs$C == "none", ]),
                 lm(I ~ D, data = costs[costs$C == "mild", ]),
                 lm(I ~ D, data = costs[costs$C == "severe", ]))
matrix(c(coef(models.I[[1]]), coef(models.I[[2]]), coef(models.I[[3]])),
       nrow = 2, ncol = 3, dimnames = list(c("(Intercept)", "D"),
                          c("none", "mild", "severe")))
              none   mild  severe
  (Intercept) 99.72 101.3  117.9
  D           299.64 549.4 790.7
```

And the standard errors are:

```
c(none = sigma(models.I[[1]]), mild = sigma(models.I[[2]]),
  severe = sigma(models.I[[3]]))
   none   mild severe
   30.4   50.2   94.2
```

We present them in the same way as they would be when printing them out by typing fitted$I.

It is important to note that this parameterisation is different from that we would get from lm by fitting a single model that includes C as an explanatory variable and as an interaction term with D.

```
single.model <- lm(I ~ D * C, data = costs)
coef(single.model)
 (Intercept)           D       Cmild     Csevere     D:Cmild    D:Csevere
       99.72      299.64        1.62       18.14      249.73       491.03
sigma(single.model)
 [1] 39.9
```

The single.model is fitted on the whole data set and has a single standard error, which is what is minimised in computing the maximum likelihood estimates. On the other hand, the mixture in models.I has one standard error per value of C, which are minimised separately in each of the calls to lm above. Hence in single.model we are assuming that all residuals are homoscedastic, while in models.I we are assuming that residuals are homoscedastic within each subset.

We can also compare the regression coefficients in models.I to those we obtain by adding the main terms ("(Intercept)" and "D") with those for the interaction terms (*i.e.*,"Cmild" and "D:Cmild") in single.model.

```
coef(single.model)["(Intercept)"] + coef(single.model)["Cmild"]
 (Intercept)
       101.3
coef(single.model)["D"] + coef(single.model)["D:Cmild"]
     D
 549.4
```

The two sets of coefficients will not be identical in the general case, even though they agree to several digits in this case.

From a more abstract point of view, we can say that the main difference between the mixture of regressions in models.I and the single regression model in single.model is to what extent information is shared across the subsets corresponding to none, mild and severe pre-conditions. In models.I there is *no pooling* of information between subsets: the parameters in each regression model are estimated using only the corresponding subset of the data. On the other hand, in single.model there is *complete pooling* in the sense that the regression model for each subset is that fitted on the whole data set.

Somewhere in between these two extremes lie Bayesian hierarchical models which employ *partial pooling* to share some information between subsets while still yielding different models for different subsets. An example are the *linear mixed effect models* illustrated in Pinheiro and Bates (2000) and implemented in the **lme4** package.

```
library(lme4)
mixmod <- lmer(I ~ D + (1|C) + D:C, data = costs)
```

The specific model we are fitting above includes one main term for D and the interaction term for D and C as fixed effects (fixef); and a random intercept (ranef) for C to allow each pre-condition to have different regression coefficient and a different variance. The maximum likelihood estimates for the associated parameters are shown below.

```
as.data.frame(ranef(mixmod))
    grpvar        term    grp condval condsd
1        C (Intercept)   none   -5.36   1.04
2        C (Intercept)   mild   -3.60   1.79
3        C (Intercept) severe    8.96   4.04
fixef(mixmod)
(Intercept)           D    D:Cmild   D:Csevere
      105.2       299.6      249.7       491.7
```

Again, adding the "(Intercept)" term and the values for the random intercepts yields 99.82, 101.57, 114.13; and adding the "D" term and "D:Cmild", "D:Csevere" yields 299.62, 549.33, 791.31. All these estimates are close to the parameter values we specified in Section 3.2. And unlike single.model, mixmod has a different standard error for each level of pre-condition, which is given by adding the standard deviations in the condsd column above to sigma(mixmod).

Which of these models should we use to estimate the parameters of I in the CGBN? Each comes with a different set of assumptions, which may or may not fit a specific data set well. Pooling becomes crucial when there are few observed samples for some configurations of the discrete parents: sharing information between different configurations produces more accurate parameter estimates for those configurations. There are also situations in which we may want to reduce the number of effective parameters to improve the stability of maximum likelihood estimates. Then we can use a single linear model like single.model and complete pooling to have a single standard error for all observations. In either case we can take these parameter estimates and replace those in fitted$I using the assignment operator as we did with ridge regression in Section 2.4.

3.4 Learning the DAG Structure: Tests and Scores

Learning the structure of a CGBN from data is largely an extension of the material in Sections 1.5 and 2.5. However, the only approaches in use in practical applications are those that use network scores to rank candidate DAGs. The most common network score is again BIC, which takes the form

$$
\begin{aligned}
\text{BIC} = \log \widehat{f}(\mathsf{A}, \mathsf{C}, \mathsf{H}, \mathsf{D}, \mathsf{I}, \mathsf{O}, \mathsf{T}) - \frac{d}{2} \log n = \\
= \left[\log \widehat{\Pr}(\mathsf{A}) - \frac{d_\mathsf{A}}{2} \log n \right] + \left[\log \widehat{\Pr}(\mathsf{C} \mid \mathsf{A}) - \frac{d_\mathsf{C}}{2} \log n \right] + \\
+ \left[\log \widehat{\Pr}(\mathsf{H} \mid \mathsf{A}) - \frac{d_\mathsf{H}}{2} \log n \right] + \left[\log \widehat{f}(\mathsf{D} \mid \mathsf{A}, \mathsf{H}) - \frac{d_\mathsf{D}}{2} \log n \right] + \\
+ \left[\log \widehat{f}(\mathsf{I} \mid \mathsf{C}, \mathsf{D}) - \frac{d_\mathsf{I}}{2} \log n \right] + \left[\log \widehat{f}(\mathsf{O} \mid \mathsf{A}) - \frac{d_\mathsf{O}}{2} \log n \right] + \\
+ \left[\log \widehat{f}(\mathsf{T} \mid \mathsf{O}, \mathsf{I}) - \frac{d_\mathsf{T}}{2} \log n \right] \quad (3.4)
\end{aligned}
$$

similarly to Equations (2.13) and (1.14). For nodes like I that are modelled with a mixture of linear regressions, the BIC term can be written as

$$
\log \widehat{f}(\mathsf{I} \mid \mathsf{C}, \mathsf{D}) - \frac{d_\mathsf{I}}{2} \log n = \sum_{c \in \mathsf{C}} \left[\log \widehat{f}(\mathsf{I} \mid \mathsf{C} = c, \mathsf{D}) - \frac{d_{\mathsf{I},c}}{2} \log n \right] \quad (3.5)
$$

by separating the contributions of the individual linear regression models for the different levels of pre-conditions $c \in \mathsf{C}$. So for instance,

$$
\log \widehat{f}(\mathsf{I} \mid \mathsf{C} = \mathtt{mild}, \mathsf{D}) = N(101 + 549\mathsf{D}, 50.2) \quad (3.6)
$$

using the parameter estimates in `fitted$I`. The number of parameters d_I is equal to the sum of the number of parameters $d_{\mathsf{I},c}$ of the individual regressions in the mixture. From the equation above, $d_{\mathsf{I},c} = 3$ for all $c \in \mathsf{C}$; each regression has one intercept, one regression coefficient for D and one standard error. Note that it is important to count the standard errors in d_I; that was not the case for GBNs because each node can only have exactly one.

The `score` function computes BIC for a CGBN by default, and the same is true for the hill-climbing implementation in `hc`. For the sake of being explicit, we can use it by passing `score = "bic-cg"` to the latter and try to learn a DAG from the `costs` data we used in the previous section.

```
learned <- hc(costs, score = "bic-cg")
modelstring(learned)
[1] "[A][C|A][H|A][O|A][I|C:H:O][D|C:I][T|I:O]"
modelstring(dag)
[1] "[A][C|A][H|A][O|A][D|A:H][I|C:D][T|I:O]"
```

If we compare the modelstring of the learned DAG with that of the true DAG from Section 3.2, we can see that D does not depend on H. The reason for this error is, interestingly, that $\log \widehat{f}(D \mid A, H)$ is equal to -Inf.

```
score(dag, costs, type = "bic-cg", by.node = TRUE)
       A         C         D        H          I         O         T
 -2099.90  -1257.21      -Inf -1019.21 -10017.55  -8673.74  -7474.70
```

This stems from the fact that D | A, H is singular: a number of standard errors are exactly equal to zero, which makes $\log \widehat{f}(D \mid A, H)$ a degenerate distribution with an infinite log-likelihood. Which would make hc return the first singular model it encounters as the best learned DAG! This is undesirable because none of the statistical properties regular models that we can find in the literature holds for singular models, making them problematic to use and interpret. As a design choice, **bnlearn** prevents that from happening by assigning a -Inf score to singular models.

3.5   Using Conditional Gaussian Bayesian Networks

To conclude this chapter, we use the local distributions of the variables in the BN to perform inference, which we have previously explored in Sections 1.6 and 2.6. Our goal is two-fold: to check that the model agrees with publicly-available information on healthcare expenditures, and to reason about the amount of taxes required to finance healthcare in a sustainable and fair manner. Unfortunately, no R package implements exact inference for CGBNs: hence we will focus on the cpquery and cpdist functions for approximate inference provided by **bnlearn**.

Firstly, we check that the average per-day expenditure for an inpatient is about £400. In order to do this, we supply to cpdist the evidence that inpatients have spent at least 1 full day at a hospital (D >= 1) and thus have caused at least the fixed-costs expenditures (I >= 100). After simulating 10^5 samples from the BN, we remove the £100 fixed costs and compute the average of I over D along with some relevant quantiles.

```
part <- cpdist(healthcare, nodes = c("I", "D"),
               evidence = (D >= 1) & (I >= 100), n = 10^5)
per.day <- (part$I - 100) / part$D
c(mean = mean(per.day), quantile(per.day, c(0.01, 0.99, 0.999)))
   mean     1%    99%  99.9%
    404    274    817    856
```

The result is indeed close to £400, and for the vast majority of people ranges from £274 to £856. As for outpatient expenditures O, there is a much wider variability as we can see below.

```
part <- cpdist(healthcare, nodes = "O", evidence = (O >= 0), n = 10^5)
summary(part$O)
   Min. 1st Qu.  Median    Mean 3rd Qu.    Max.
     17      66     171     174     205     509
```

However, the range and the distribution of the simulated values are realistic. In the lower quartile, we have people who go their GP up to two times a year (which costs on average £37.5 per visit in 2018) and require up to two prescriptions (at £8 each). On the other hand, referrals cost on average £180 each; going to Accidents & Emergency (A&E) can cost anywhere between £45 and £400; ambulance trip costs on average £252 or £192, depending on whether they result in a trip to A&E or not. (And in Chapter 5 you can find out how long you have to wait once you get there!) Hence it is easy to see how expenditures can increase all the way to £509 in the event of some serious accident or follow-up visits.

We can then check D and C. For the former, the average should be between 4.5 and 5. For the latter, around 14% of people under 40 should have a pre-existing condition; since A == "young" identifies people under 30 we expect an even smaller number. Similarly, we expect about 60% of old people to have some sort of condition.

```
part <- cpdist(healthcare, nodes = "D", evidence = (H == "any"),
        n = 10^5)
c(mean = mean(part$D), quantile(part$D, c(0.01, 0.99)))
  mean    1%   99%
 4.567 0.238 9.836
cpquery(healthcare, event = (C %in% c("mild", "severe")),
      evidence = (A == "young"), n = 10^5)
 [1] 0.118
cpquery(healthcare, event = (C %in% c("mild", "severe")),
      evidence = (A == "old"), n = 10^5)
 [1] 0.582
```

This leads to our assessment of the amount of tax needed to cover these healthcare expenditures, which we can again evaluate by simulation using cpdist to generate samples for I, O and T. Note that we require both that $I \geqslant 0$ and that $O \geqslant 0$ to ensure that results are realistic: both I and O are modelled using normal random variables and therefore they can technically assume negative values, even though in our BN they do so with small probabilities given their variances.

```
part <- cpdist(healthcare, nodes = c("I", "O", "T"),
        evidence = (I >= 0) & (O >= 0), n = 10^5)
summary(part$T)
   Min. 1st Qu.  Median    Mean 3rd Qu.    Max.
    144     313     416     856     629    9438
```

The first quartile and the median are, in fact, not very different from the value of the Immigration Health Surcharge (£300 for people between 18 and 30, £400 for everybody else) that the government charged people moving to the UK from outside of the European Economic Area in 2018. On average, the average amount of tax money and the average expenditures are as follows.

```
finances <- c(mean.tax = mean(part$T),
              mean.expenditure = mean(part$I + part$O),
              surplus = mean(part$T) - mean(part$I + part$O))
finances
         mean.tax mean.expenditure         surplus
              856              716             140
```

This means £71.33 per month of taxes, of which £59.7 are used to cover the expenditures and the rest can be reinvested or used to waive the tax for low-income individuals and children. As we would expect, if we actually charge £856 of taxes to everybody, most young people cost less money than they pay in, and old people often cost more money than they pay in.

```
cpquery(healthcare, event = (T <= finances["mean.tax"]),
        evidence = (A == "young"))
 [1] 0.99
cpquery(healthcare, event = (T > finances["mean.tax"]),
        evidence = (A == "old"))
 [1] 0.396
```

Will this amount of tax allow a sustainable healthcare system in the future? One trend to take into consideration is that the UK population is getting older over time. We can simulate how that would affect expenditures by re-running the simulations above after changing the distribution of A.

```
new.A.prob <- array(c(0.30, 0.40, 0.30), dim = 3,
                    dimnames = list(A = A.lv))
new.A.prob
 A
 young adult   old
   0.3   0.4   0.3
healthcare$A <- new.A.prob
```

Now we have 10% fewer young people, 5% fewer adult people and 15% more old people.

```
part <- cpdist(healthcare, nodes = c("I", "O"),
        evidence = (I >= 0) & (O >= 0), n = 10^5)
finances["mean.tax"] - mean(part$I + part$O)
 mean.tax
     9.62
```

The tax surplus after covering healthcare expenditures decreases from £140 to about £10, which means we would effectively run a balanced budget. Another factor to consider is that the incidence of various conditions may vary with time; even a small increase in severe conditions could make expenditures overrun the budget. An example:

```
new.C.prob <- array(c(0.88, 0.10, 0.02, 0.70, 0.22, 0.08, 0.41,
                      0.51, 0.08), dim = c(3, 3),
                dimnames = list(C = C.lv, A = A.lv))
new.C.prob - C.prob
        A
C          young adult   old
   none       0 -0.05 -0.01
   mild       0  0.02 -0.02
   severe     0  0.03  0.03
healthcare$C <- new.C.prob
part <- cpdist(healthcare, nodes = c("I", "O"),
         evidence = (I >= 0) & (O >= 0), n = 10^5)
finances["mean.tax"] - mean(part$I + part$O)
 mean.tax
   -4.06
```

This would force us to increase taxes, or (as has happened in most countries) to reduce the number of days D of hospital stays by changing treatment protocols. Consider, for example, the effect of an average reduction of D by 1 day for patients with severe conditions.

```
new.D.coef <- list(coef = array(c(0, 0, 0, 1, 4, 6), dim = c(1, 6),
                         dimnames = list("(Intercept)", NULL)),
                 sd = c(0, 0, 0, 0.5, 1, 1.5))
healthcare$D <- new.D.coef
part <- cpdist(healthcare, nodes = c("I", "O"),
         evidence = (I >= 0) & (O >= 0), n = 10^5)
finances["mean.tax"] - mean(part$I + part$O)
 mean.tax
    40.4
```

This change alone is enough to bring back a surplus of £40, making healthcare sustainable once more.

3.6 Further Reading

CGBNs are rarely covered in books on graphical models compared to discrete BNs and GBNs. They are explored, however, in Koller and Friedman (2009, Section 5.5.1 and Chapter 14), Koski and Noble (2009, Section 8.9) and Kjærluff and Madsen (2013, Sections 4.1.2 and 5.1.2).

Exercises

Exercise 3.1 *Consider the expression for BIC in Equation (3.4). What are the values of d_A, d_C, d_H, d_D, d_I, d_O and d_T? Or, equivalently, how many parameters does each local distribution have? How many (fewer) parameters would the corresponding CGBN fitted with complete pooling have?*

Exercise 3.2 *The* **deal** *package contains a data set called* ksl, *which describes the health and social characteristics of a sample of Danish 70 year olds. It includes the following variables:* FEV *(Forced ejection volume),* Kol *(Cholesterol),* Hyp *(Hypertension),* logBMI *(Logarithm of Body Mass Index),* Smok *(Smoking),* Alc *(Alcohol consumption),* Work *(Working),* Sex *an* Year.

    1. Load the ksl *data, ensure that all variables are either* "factor" *or* "numeric", *and use* hc *to learn the structure of a BN.*

    2. Learn the parameters of the BN.

    3. Use cpquery *to compute the probability of having hypertension in the general population, in individuals with cholesterol higher than 750, and in individuals with log-BMI higher than 3.5. Are these probabilities comparable?*

    4. Use cpdist *to simulate the joint distribution of cholesterol and log-BMI for individuals that smoke but have no hypertension. Are these two variable correlated?*

Exercise 3.3 *One of the first examples of mixed-effects models in Pinheiro and Bates (2000) describes a randomised-block experiment which recorded the effort (E) required by each of nine different subjects (S) to arise from each of four types of stools (T).*

    1. Load the ergoStool *data for this experiment from the* **nlme** *package, and rename the variables to* E, S, T.

    2. Fit a mixed-effects model in which T *is a fixed effect,* S *is a random intercept effect and* E *is the response variable; and the corresponding classic linear model in which* T *and* S *are the explanatory variables.*

    3. Learn the parameters of the CGBN with DAG $S \rightarrow E \leftarrow T$ *from* ergoStool.

    4. Comment on the differences and the similarities in the parameterisations of these three models.

4

Time Series: Dynamic Bayesian Networks

All the examples we have considered up to this point are *static*: each statistical individual is measured just once at some point in time. Many interesting problems, however, are *dynamic* in nature and the analysis of the data they generate focuses on how individuals evolve over time. BNs presented in previous chapters can be extended into *dynamic BNs* (DBNs) to perform this kind of analysis, which will be the topic of this chapter.

4.1 Introductory Example: Domotics

Suppose we have a house in which we want to install a microcontroller that opens and closes windows to improve air quality through passive ventilation depending on outside conditions. The microcontroller will be connected to a number of sensors which will be placed both inside and outside the house and which will take various environmental measurements every 10 minutes. Using these measurements, the microcontroller will use a simple DBN to predict the air quality inside the house in 10 minutes time and open/close the windows to keep the air fresh and a comfortable inside temperature. This kind of home automation is called *domotics*, and represents a greener alternative than using air conditioning given in suitably mild climates.

We can think of indoor air quality as a combination of *temperature* and *stuffiness*, which we can define as the combination of humidity and the concentration of CO_2. As for the outside conditions, we care mainly about the *outside temperature*. If windows are open, conditions inside the house will approach those outside due to air exchange. If the inside is cooler than the outside, opening the windows will increase the inside temperature. On the other hand, if the inside is hotter than the outside the inside temperature will decrease. Regardless of the temperature, opening the windows will let fresh air in and let stuffy air out thus improving the inside conditions.

All these changes, however, take time: windows can only let a limited volume of air in and out at once.

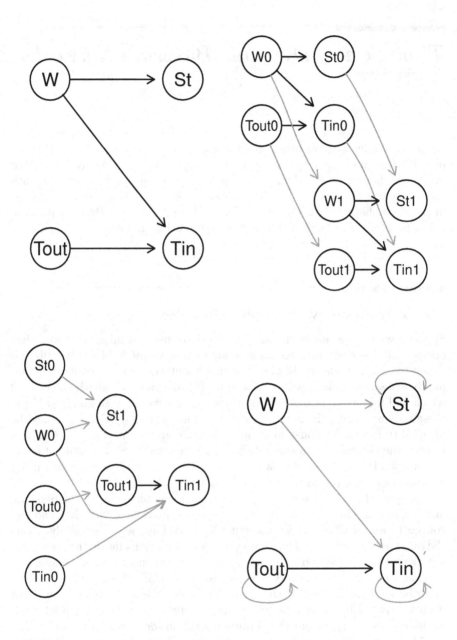

Figure 4.1
DAGs for a static BN (top left), a naive DBN (top right), a more parsimonious
DBN (bottom left) and the equivalent rolled-up DBN (bottom right). Arcs
that span different time points are shown in grey.

4.2 Graphical Representation

A DBN models time explicitly by creating duplicates of the variables at each time point they are measured at. From the description above, our variables of interest are stuffiness (St), inside temperature (Tin), outside temperature (Tout) and a binary variable (W) that models whether windows are open or not. If we disregard time, we could model them with a DAG like that in the top-left panel of Figure 4.1.

Starting from this DAG, we can argue from common sense that St, Tin and Tout change smoothly over time; so their values at a given time (say, t_1) should depend on the respective values 10 minutes earlier (say, t_0). The same is true for W, as we are not expecting to constantly open and close the windows every 10 minutes. We could model this intuition by taking the previous network, making a copy of the DAG for t_1 (while the original models the variables at t_0) and connecting each node in t_0 with the corresponding node in t_1 to obtain the DAG in the top-right panel. This new network encodes two assumptions: that the dependence structure between the nodes is the same at t_0 and t_1; and that variables at t_1 depend on those at t_0 but not on those at earlier time points.[1] Implicitly, it also assumes that t_0 and t_1 are not instants, but rather averages over periods of time: if that were not the case it would be problematic to draw arcs between nodes in the same time point because that would imply they can affect each other without any time passing.

However, this second network has an important limitation: the marginal distributions of the nodes in t_0 are not required to be identical to the corresponding distributions in t_1. As a result, the network has a higher number of parameters and it will not be necessarily consistent when those parameters are estimated from data. (The same observation may have very different marginal probabilities at t_0 and t_1!) So, instead of constructing a network modelling t_0, t_1 and the transition between t_0 and t_1, we prefer to assume that nodes in t_0 have the same distribution as those in t_1 and focus on modelling the transition between t_0 and t_1.

This leads to the DBN in the bottom-left panel of Figure 4.1. Since we are not interested in modelling t_0 in detail, we dropped all the arcs between the nodes in that time point and we basically treat them as fixed. The remaining arcs connect nodes in t_0 to nodes in t_1, to model changes over time, and pairs of nodes in t_1. To make the DBN simpler, we also dropped the node W1 in favour of W0 so that we only consider the effect of the windows being open or closed at time t_0 to influence Tin1, Tout1 and St1. Considering that the aim of

---

[1]In other word, the network in the top-right panel represents a first-order Markov process. Furthermore, that in the bottom-left panel is also homogeneous. DBNs can represent higher-order Markov processes using even larger and more complex networks, but we will not explore this topic further.

the DBN is to predict Tin1 and St1 from the nodes in t_0, neither simplification is problematic.

This final network can be represented in a more compact way as in the bottom-right panel by merging back nodes across time points in what is called a *rolled-up* graph. Loops (that is, arcs from a node to itself) represent arcs from a node in t_0 to the corresponding node in t_1 (say, St0 → St1). Other arcs are taken to go from a node in t_0 to a node in t_1, with the exception of Tout → Tin which represents Tout1 → Tin1 (and is therefore drawn in a different color).[2] Note that rolled-up graphs are generally not DAGs because they may contain both loops and cycles!

4.3   Probabilistic Representation

DBNs do not come with any specific assumptions on the form of the probability distributions that can be used for the local distributions: discrete BNs, GBNs, CGBNS as well as advanced BNs such as those we will cover in Chapter 5 can all be adapted into DBNs. For simplicity, we choose to formulate this example as a discrete BN like in Chapter 1.

First of all, we categorise temperatures as *comfortable* (18°–24°C), *cold* (less than 18°C) and *hot* (more than 24°C). We assume that outside temperatures are for the most part comfortable in Tout0, and that Tout1 has a 0.80 probability of being the same category as Tout0.[3]

```
T.lv <- c("<18", "18-24", ">24")
Tout0.prob <- array(c(0.20, 0.70, 0.10), dim = 3,
                dimnames = list(Tout0 = T.lv))
Tout1.prob <- array(c(0.80, 0.19, 0.01, 0.10, 0.80, 0.10,
                    0.01, 0.19, 0.80), dim = c(3, 3),
                dimnames = list(Tout1 = T.lv, Tout0 = T.lv))
Tout1.prob
          Tout0
Tout1    <18 18-24  >24
  <18    0.80  0.1 0.01
  18-24  0.19  0.8 0.19
  >24    0.01  0.1 0.80
```

As for W0, we just say that windows are open or closed with probability

---

[2]Technically, rolled-up graphs have originally been defined for DBN without instantaneous dependencies, so making this distinction is an extension of the original definition.

[3]These would be sensible values for early summer in the UK, the effects of global warming notwithstanding.

0.5. As mentioned earlier, all inference on this DBN will be conditional on the value of W0 which means this distribution will not be used in practice.

```
W.lv <- c("open", "closed")
W0.prob <- array(c(0.5, 0.5), dim = 2, dimnames = list(W0 = W.lv))
```

For simplicity, we then categorise stuffiness as *low* or *high* and we again choose a uniform distribution for St0. We assume that St1 will take the same value as St0 with high probability, but with a lower probability of observing St1 == "low" if windows are closed than if they are open.

```
St.lv <- c("low", "high")
St0.prob <- array(c(0.50, 0.50), dim = 2, dimnames = list(St0 = St.lv))
St1.prob <- array(c(0.90, 0.10, 0.70, 0.30, 0.70,
                    0.30, 0.10, 0.90), dim = c(2, 2, 2),
             dimnames = list(St1 = St.lv, St0 = St.lv, W0 = W.lv))
St1.prob
, , W0 = open

      St0
St1    low high
  low  0.9  0.7
  high 0.1  0.3

, , W0 = closed

      St0
St1    low high
  low  0.7  0.1
  high 0.3  0.9
```

Finally, we choose the distribution of the inside temperature Tin0 to give a high probability of a comfortable temperature. Then Tin1 will have a high probability of having the same value as Tin0, and at the same time it will have a smaller (but still markedly different from zero) probability of moving up or down by one category depending on whether Tout1 is higher or lower than Tin1. This probability will be larger if windows are open than if they are closed. We also assume that it is not possible to move from less than $18°C$ to more than $24°C$ or vice versa between t_0 and t_1, allowing only smooth changes in temperature.

```
Tin0.prob <- array(c(0.10, 0.85, 0.05), dim = 3,
               dimnames = list(Tin0 = T.lv))
Tin1.prob <- array(c(
         # W0 = "open", Tin0 = "<18"
         0.875, 0.125, 0, 0.075, 0.9, 0.025, 0.075, 0.7, 0.225,
         # W0 = "closed", Tin0 = "<18"
         0.875, 0.125, 0, 0.475, 0.5, 0.025, 0.025, 0.65, 0.325,
```

```
        # W0 = "open", Tin0 = "18-24"
        0.475, 0.525, 0, 0.075, 0.8, 0.125, 0, 0.875, 0.125,
        # W0 = "closed", Tin0 = "18-24"
        0.075, 0.9, 0.025, 0, 0.875, 0.125, 0, 0.475, 0.525,
        # W0 = "open", Tin0 = ">24"
        0.15, 0.725, 0.125, 0, 0.475, 0.525, 0, 0.475, 0.525,
        # W0 = "closed", Tin0 = ">24"
        0, 0.125, 0.875, 0, 0.075, 0.925, 0, 0.175, 0.825),
    dim = c(3, 3, 2, 3),
    dimnames = list(Tin1 = T.lv, Tout1 = T.lv, W0 = W.lv, Tin0 = T.lv))
```

Having created all the conditional probability tables, we can combine then with the DAG in the bottom-left panel of Figure 4.1 to create the bn.fit object encoding the DBN.

```
dag <- model2network(paste0("[W0][St0][Tout0][Tin0][St1|St0:W0]",
            "[Tout1|Tout0][Tin1|Tin0:W0:Tout1]"))
cpt <- list(Tout0 = Tout0.prob, Tout1 = Tout1.prob, W0 = W0.prob,
        Tin0 = Tin0.prob, Tin1 = Tin1.prob, St0 = St0.prob,
        St1 = St1.prob)
dbn <- custom.fit(dag, cpt)
```

Despite all the simplifications we have made and the fairly simple distributions we have chosen, dbn still has a larger number of parameters than other the examples we have considered previously.

```
nparams(dbn)
  [1] 52
```

This is what motivates us to make those simplifications in the first place: it is very easy for a DBN to require a large number of parameters. Among other problems, it becomes increasingly difficult to compile or to estimate well large conditional probability tables such as that of Tin1.

4.4 Learning a Dynamic Bayesian Network

Estimating the parameters of a DBN does not require any statistical method that is specific to DBNs: they make, after all, the same assumptions about the distributions of the nodes as the other types of BNs we explore in this book. Hence we refer the reader to Section 6.5.2 for a general overview of parameter learning and to Section 1.4 for estimators that are specific to discrete BNs.

Learning the structure of the DAG that underlies a DBN, however, is different in a few specific ways from the general case (which we will cover in Section 6.5.1) and from what we saw in Section 1.5 for discrete BNs. In order to capture the evolution of the variables over time, some nodes are measured at t_0 and some at t_1.

```
t0.nodes <- c("W0", "St0", "Tout0", "Tin0")
t1.nodes <- c("St1", "Tout1", "Tin1")
```

The two groups are treated differently when learning the DAG structure. Nodes in t_0 are not linked to each other by arcs because we do not care to model their probabilistic dependencies given how we will use the DBN. This means that we should *blacklist* all possible arcs between nodes in t_0 in order to make sure they are not considered during structure learning.

```
bl <- set2blacklist(t0.nodes)
```

We should also blacklist all the arcs pointing from nodes in t_1 to nodes in t_0, as those arcs would represent a variables in the "past" depending on a variable in the "future". This would make their interpretation, and the interpretation of any inference in the next section, much less intuitive as it would go against our causal understanding of the world (more in Section 6.7).

```
bl <- rbind(bl, tiers2blacklist(list(t0.nodes, t1.nodes)))
head(bl)
         from    to
  [1,]  "St0"   "W0"
  [2,]  "Tout0" "W0"
  [3,]  "Tin0"  "W0"
  [4,]  "W0"    "St0"
  [5,]  "Tout0" "St0"
  [6,]  "Tin0"  "St0"
```

The blacklist produced by `set2blacklist` and `tiers2blacklist` is a matrix with the same structure as that returned by the `arcs` function. `set2blacklist` takes a set of nodes and produces a blacklist containing all possible arcs connecting them to each other; `tiers2blacklist` takes a list of sets of nodes (the *tiers*) and returns a blacklist containing all the arcs pointing from any node in one tier to any node in a preceding tier. Note that blacklisting an arc between two nodes (say, St1 → St0) does not blacklist the corresponding arc in the opposite direction (St0 → St1): the algorithm we use for structure learning is free to include it or not in the DAG depending on whether it is well supported by the data.

Having a complete list of the arcs that should not be included in the DAG we will learn from data, we can pass it to (for instance) hc via the blacklist argument to learn the DBN in the bottom-left panel of Figure 4.1. For this purpose, we loaded a data set of 2000 observations in the domotics object.

```
dbn.hc <- hc(domotics, blacklist = bl)
all.equal(dag, dbn.hc)
  [1] TRUE
```

Finally, we can take the learned DAG and fit the parameters with bn.fit from the same data before using the fitted DBN for inference.

```
dbn.fit <- bn.fit(dbn.hc, domotics)
```

4.5 Using Dynamic Bayesian Networks

In Section 4.1, we introduced this example stating that the DBN's main purpose is to *predict* the air quality inside the house 10 minutes in the future in order to decide whether to open or close the windows. Therefore, and unlike the BNs in previous chapters, we will use this DBN for prediction rather than answering arbitrary queries as we did in Sections 1.6.2 and 2.6. These two inference tasks are not unrelated: in discrete BN, whether static or dynamic, the value that we predict a variable will take will be that with the highest probability given the variables we observe. Hence, we would like to predict the air quality by predicting the values of St1 and Tin1 as

$$\{\widehat{St1}, \widehat{Tin1}\} = \underset{St1,Tin1}{\operatorname{argmax}} \Pr(St1, Tin1 \mid St0, Tin0, Tout0, W0)$$

for both W0 == "open" and W0 == "closed": we can then choose whether to open or close the windows depending on which value of W0 gives the highest probability that St1 == "low" and Tin1 == "18-24". For instance, consider the following scenario: while temperature is comfortable inside and cold outside, air is stuffy at time t_0. The probabilities we estimate with cpquery are

```
cpquery(dbn.fit, event = (St1 == "low") & (Tin1 == "18-24"),
    evidence = (St0 == "high") & (Tin0 == "18-24") &
               (Tout0 == "<18") & (W0 == "closed"))
 [1] 0.0846
cpquery(dbn.fit, event = (St1 == "low") & (Tin1 == "18-24"),
    evidence = (St0 == "high") & (Tin0 == "18-24") &
               (Tout0 == "<18") & (W0 == "open"))
 [1] 0.42
```

which suggest that opening windows is preferable to keeping them closed. Even though the probability of actually getting good air quality inside the house is below 0.5 even if we open them! We can make predictions even when we do not know the values of all the variables at time t_0 by omitting them from the evidence as we do with Tout0 below.

```
cpquery(dbn.fit, event = (St1 == "low") & (Tin1 == "18-24"),
    evidence = (St0 == "high") & (Tin0 == "18-24") & (W0 == "closed"))
 [1] 0.0984
```

```
cpquery(dbn.fit, event = (St1 == "low") & (Tin1 == "18-24"),
  evidence = (St0 == "high") & (Tin0 == "18-24") & (W0 == "open"))
 [1] 0.516
```

The result is the probability of St1 == "low" and Tin1 == "18-24" given the variables we condition on, averaged over the possible values of Tout0.

bnlearn provides a predict function that can automate this process to some extent. In order to use it, we first create an artificial observation with the conditions we observe at time t_0.

```
evidence <- data.frame(St0 = factor("high", levels = St.lv),
                Tin0 = factor("18-24", levels = T.lv),
                Tout0 = factor("<18", levels = T.lv),
                W0 = factor("open", levels = W.lv))
```

Then we use this artificial observation to predict Tin1, fill in the value we predict and finally predict the value of St1.

```
predict(dbn.fit, data = evidence , node = "Tin1", method = "bayes-lw")
 [1] 18-24
 Levels: <18 18-24 >24
evidence$Tin1 <- factor("18-24", levels = T.lv)
predict(dbn.fit, data = evidence , node = "St1", method = "bayes-lw")
 [1] low
 Levels: low high
```

Either approach suggests that opening windows should be preferable to keeping them closed, so the DBN will tell the microcontroller to open them for the next 10 minutes.

4.6 Plotting Dynamic Bayesian Networks

Like structure learning and inference, plotting the DAGs that underlie DBNs should also take into account the temporal ordering of the variables. By convention, these DAGs are printed from left to right, with the nodes in t_0 aligned on the left and the nodes in t_1 grouped on the right. If we were considering additional time points (t_2, t_3, etc.), the plots would show the nodes in each of time point grouped together and each group to the right of the previous one. This is how the DAGs in the first three panels of Figure 4.1 are laid out.

graphviz.plot has an argument called groups that allows the user to specify how nodes should be clustered together when drawing the DAG, but it does not have an argument to make the dot layout to layer the nodes left-to-right instead of top-to-bottom. Hence we will combine graphviz.plot with **Rgraphviz** to achieve the desired result.

As we did in Section 1.7.1, we first call `graphviz.plot` to get a graph object that we can modify with the functions in **Rgraphviz**. This time, however, we will specify `render = FALSE` to prevent `graphviz.plot` from actually drawing a plot we are not interested in.

```
gR <- graphviz.plot(dag, render = FALSE)
```

We then group the nodes in the DAG into `t0.nodes` and `t1.nodes`, and for each of those groups we create a subgraph with the `subGraph` function from **Rgraphviz**.

```
t0.nodes <- c("W0", "St0", "Tout0", "Tin0")
t1.nodes <- c("St1", "Tout1", "Tin1")
sg0 <- list(graph = subGraph(t0.nodes, gR), cluster = TRUE)
sg1 <- list(graph = subGraph(t1.nodes, gR), cluster = TRUE)
```

Passing these subgraphs along with `rankdir = "LR"` (where `rankdir` is an optional argument of the `dot` layout and `"LR"` stands for left-to-right) to `layoutGraph` will make it draw the subgraphs from left to right and subsequently add arcs across subgraphs.

```
gR <- layoutGraph(gR, attrs = list(graph = list(rankdir = "LR")),
       subGList = list(sg0, sg1))
```

Finally, changing the color of the arcs that connect the nodes in `t0.nodes` to those in `t1.nodes` will produce the plot in the bottom-left panel of Figure 4.1.

```
cross <- c("St0~St1", "Tin0~Tin1", "Tout0~Tout1", "W0~St1", "W0~Tin1")
edgeRenderInfo(gR)$col[cross] = "darkgrey"
renderGraph(gR)
```

4.7   Further Reading

Several books on graphical models devote some space specifically to DBNs: among them Korb and Nicholson (2011) in Section 4.5, Koller and Friedman (2009) in Section 6.2.2, Kjærluff and Madsen (2013) in Section 4.4 and Sucar (2015) in Chapter 9.

DBNs are a useful tool in modelling dynamic systems in a more general machine learning setting, and are covered as such in Russell and Norvig (2009) in Section 15.5 and in Murphy (2012). They have also been extended beyond the basic formulation to model continuous variables and non-homogeneous Markov processes; for a recent review see Scutari (2020).

Exercises

Exercise 4.1 *Consider the networks in Figure 4.1.*

1. How many parameters have the networks, and the local distributions associated with the individual nodes, in the top-left, top-right and bottom-left panels?

2. Extend the network in the bottom-left panel to model as second time point t_2 in addition to t_0 and t_1. How many additional parameters does that require?

3. Finally, make the nodes in t_2 dependent on the nodes in t_0 in addition to those in t_1. How many additional parameters does that require?

Exercise 4.2 *Consider again the DBN in the bottom-left panel of Figure 4.1.*

1. Extend the network to model t_2 as in point 2 of Exercise 4.1, and create the bn object encoding it. Call the new nodes St2*,* Tin2 *and* Tout2.

2. Use the conditional probabilities from Section 4.3 to create a bn.fit *object from the* bn *object in the previous point.*

3. Use cpquery *to compute the probability that* Tin2 *is equal to* "18-24" *and* St2 *is* "low" *given that* Tin0 *is* "18-24" *and* St0 *is* "high" *when windows are either open or closed. What would you expect compared to the similar query we performed in Section 4.5?*

Exercise 4.3 *Consider the* eg1.DSE.data *in the* **dse** *package, and in particular the three output series M1, GDP and CPI indexes for Canada between 1961 and 1991.*

1. Load the data and reformat them to have variables MI_0, GDPl2_0, CPI_0, MI_1, GDPl2_1, CPI_1.

2. Learn the structure of DBN using the appropriate blacklists.

3. Learn the parameters of the DBN modelling it as GBN. What kind of model do you get as a result?

4. Revisit the first two points to learn the structure of a DBN spanning t_0, t_1 and t_2.

5

More Complex Cases: General Bayesian Networks

In this chapter we will conclude our exploration of BNs, moving to the more general case in which each variable in the data is modelled with the random variable that best suits it rather than limiting ourselves to multinomial and normal distributions. For this purpose, we will use the Stan (Carpenter et al., 2017) MCMC sampler through its interface **rstan** (Stan Development Team, 2020b).

5.1 Introductory Example: A&E Waiting Times

Suppose that we are interested in estimating the waiting times in the Accidents & Emergency (A&E) department of a hospital. Much information is publicly available on the subject, since this is one of the key metrics A&E departments are evaluated on. For instance, the House of Commons[1] and NHS England[2] regularly report the relevant statistics on this subject: we will use them as a source of expert knowledge in constructing our BN.

Patients that present themselves to A&E are prioritised based on the severity of their symptoms; this process is called *triage*. Clearly, some patients arrive in critical condition and need immediate attention; some can wait for a short time before treatment is administered; while others need little or no medical treatment at all. Two important factors that may determine which category patients fall in are the type of *incident* (I) they were involved in and their *age* (A), since older people are physically more fragile and recover more slowly. We take these two variables to largely determine the *trauma score* (S), which is defined by the Smart Incident Command System triage system on a scale from 0 to 12. This is a vastly simplified characterisation, which we choose for the sake of the example: a real triage process takes into account other information such as co-morbidities (diabetes, cancer, etc.) and many other risk factors (obesity, high blood pressure, etc.).

---

[1]https://commonslibrary.parliament.uk/research-briefings/cbp-7281

[2]https://www.england.nhs.uk/statistics/statistical-work-areas/ae-waiting-times-and-activity

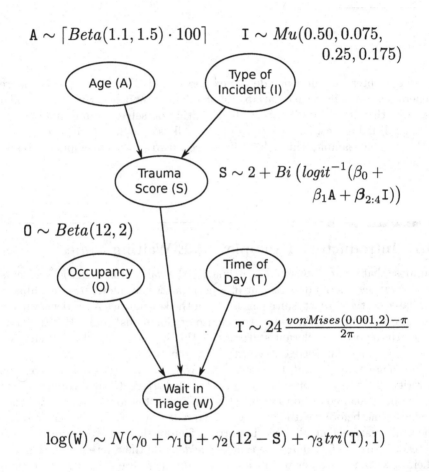

$A \sim \lceil Beta(1.1, 1.5) \cdot 100 \rceil$ $I \sim Mu(0.50, 0.075,$
$0.25, 0.175)$

Age (A)

Type of
Incident (I)

Trauma
Score (S) $S \sim 2 + Bi\left(logit^{-1}(\beta_0 +$
$\beta_1 A + \beta_{2:4} I)\right)$

$O \sim Beta(12, 2)$

Occupancy
(O)

Time of
Day (T)

$T \sim 24 \frac{vonMises(0.001,2) - \pi}{2\pi}$

Wait in
Triage (W)

$\log(W) \sim N(\gamma_0 + \gamma_1 O + \gamma_2(12 - S) + \gamma_3 tri(T), 1)$

Figure 5.1
The DAG and the local probability distributions for the A&E waiting times
BN.

The trauma score is indicative of how urgently the patient requires treatment, so it is strongly indicative of how long the patient is likely to *wait in triage* (W). However, both the hospital's bed *occupancy rate* (O) and the *time of the day* (T) may prolong waiting times. Higher occupancy rates make it more difficult to find a free bed in the hospital to administer medical care. And some times of the day are busier than others, with more people presenting themselves at the A&E during the day than during the night.

5.2 Graphical and Probabilistic Representation

From the description above we can infer the following relationships:

$$\{I, A\} \rightarrow S, \qquad\qquad \{S, O, T\} \rightarrow W,$$

which are shown in DAG form in Figure 5.1. An important characteristic of this DAG is that it divides our BN in two separate submodels: one for assigning the trauma score ($\{I, A\} \rightarrow S$) and one for the waiting time itself ($\{S, O, T\} \rightarrow W$). Once the trauma score is determined, the waiting time is independent from both the incident type and the patient's age: it is easy to verify this is the case testing d-separation of some nodes in different submodels with the dsep function.

```
dag = model2network("[I][A][S|I:A][O][B][W|S:O:B]")
dsep(dag, x = "A", y = "W", z = "S")
 [1] TRUE
dsep(dag, x = "I", y = "O", z = "S")
 [1] TRUE
```

Since we are not limiting ourselves to multinomial or normal distributions, we have more freedom in modelling available expert knowledge in a realistic way. For the type of incident, we consider four categories and we model them with a multinomial random variable.

$$I = \begin{cases} \text{domestic incident (domestic)} & \text{with probability} \quad 0.50 \\ \text{road traffic incident (road)} & \text{with probability} \quad 0.075 \\ \text{work incident (work)} & \text{with probability} \quad 0.25 \\ \text{other incident (other)} & \text{with probability} \quad 0.175 \end{cases}.$$

In most NHS documents, the age of patients is described in 5-years brackets for ages 0–100, which does not make for a parsimonious model as it would require 20 parameters. Instead, we choose a beta distribution multiplied by 100 and rounded to unit values:

$$A \sim \lceil Beta(1.1, 1.5) \cdot 100 \rceil.$$

This approach preserves the main features of the age distribution of people who attend A&E: the probability of attending A&E it is relatively flat between the ages of 30 and 70; it has a small peak for children aged 0 to 5, and it is larger for people in their 20s than it is for older adults; it gives increasing large probabilities as age increases beyond 70; 25% of patients are younger than 20 and 20% are older than 65. Combined with the absolute frequencies of different ages, as they can be gleaned from a population pyramid plot,[3] we obtain a distribution like that the top panel of Figure 5.2. It is easy to check that the 0.25 and 0.80 quantiles roughly match the information above.

```
round(100 * qbeta(0.25, 1.1, 1.5))
 [1] 20
round(100 * qbeta(0.80, 1.1, 1.5))
 [1] 68
```

The trauma score is an integer number between 0 and 12, but scores between 0 and 2 are almost never used in practice. Hence S can be naturally modelled as $S \sim 2 + Bi(p, 10)$ where the Binomial probability of each point is determined by a logistic regression:

$$\log \left(\frac{p}{1-p} \right) = \beta_0 + \beta_1 A + \beta_2 \mathbb{1}(I = \text{road}) + \beta_3 \mathbb{1}(I = \text{work}) + \beta_4 \mathbb{1}(I = \text{other}).$$

$$(5.1)$$

In other words, we model the log-odds-ratio as a baseline (β_0, determined using domestic incidents since they are the most common) that increases with age ($\beta_1 A$) and that is adjusted for the average severity of road incidents ($\beta_2 \mathbb{1}(I = \text{road})$), work incidents ($\beta_3 \mathbb{1}(I = \text{work})$) and other kinds of incidents ($\beta_4 \mathbb{1}(I = \text{other})$) relative to domestic incidents. Trauma scores are grouped in four brackets, denoted by color codes[4]: *black* (0–2, beyond help), *red* (3–10, need immediate attention), *yellow* (10–11, can wait for a short time) and *green* (12, treatment can be delayed). Hence we set the baseline $\beta_0 = 7$; we decrease it by 1 point for each 20 years of age with $\beta_1 = -0.05$; and we further decrease it by $\beta_3 = -4$, $\beta_4 = -3$, $\beta_5 = -1$ for the various types of incidents.

The occupancy rate is a proportion, so we can model it with a beta distribution. we know from NHS statistics that the average is ≈ 0.90, and that the density should be concentrated between 0.80 and 0.99. This leads to an $O \sim Beta(12, 2)$, shown in the middle panel of Figure 5.2: it has expected value $E(O) \approx 0.89$ and $Pr(O \in [0.80, 0.99]) \approx 0.87$.

As for the time of the day T, we need a periodic function that can express the cyclic frequency of patients' arrivals. One way to do this is to use a density function defined over a circle, which ensures continuity as we transition from one day to the next. One such distribution is the *von Mises* distribution, which is defined over $[-\pi, \pi]$ and has two parameters μ (where the peak is in each

---

[3]*i.e.,*   https://www.ons.gov.uk/peoplepopulationandcommunity/populationandmigration/
populationestimates/articles/ukpopulationpyramidinteractive/2020-01-08
[4]https://en.wikipedia.org/wiki/Triage#United_Kingdom

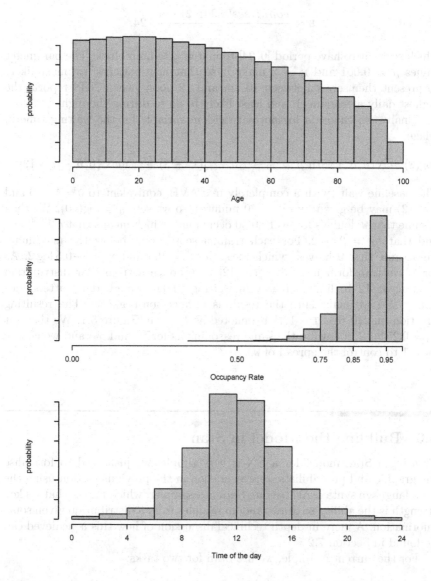

Figure 5.2
Distributions of Age (A), Occupancy (O) and arrivals for each Time (T) of the day on a 24-hour clock.

period) and κ (how sharp the peak is). Hence we define T as

$$\mathsf{B} \sim \frac{vonMises(0.001, 2) - \pi}{2\pi} \cdot 24,$$

which scales it to have period $[0, 24]$ to match a 24-hour clock. The parameter values $\mu = 0.001$ and $\kappa = 2$ make it so that new patients are most likely to present themselves between 10 am and 12 noon (when NHS reports the highest daily attendance), and least likely to do so during the night.

Finally, we choose a log-normal random variable for the waiting time in triage:

$$\log(\mathsf{W}) \sim N(\mu, \sigma^2) \text{ with } \mu = \gamma_0 + \gamma_1 0 + \gamma_2(12 - \mathsf{S}) + \gamma_3 \max\{0, 6 - |\mathsf{T} - 12|\}.$$

The baseline wait (with a completely free A&E, equivalent to $\mathsf{O} = \mathsf{T} = 0$ and $\mathsf{S} = 12$) may be given by $e^{\gamma_0} = 20$ minutes, so we set $\gamma_0 = \log(20)$. We then assume that W doubles for every 0.50 occupancy, which means that $e^{\gamma_1 0.50} = 2$ and that $\gamma_1 = 2\log 2$. For each trauma score point below the maximum, we assume that W halves, which means $e^{\gamma_2 2} = 0.5$ and $\gamma_2 = -0.5\log 2$. As for T, we transform it as $6 - |\mathsf{T} - 12|$ to produce a triangular distribution peaking at 12 to link increases in waiting times to high daily attendance hours; and we make sure the result is always non-negative. The resulting function $\max\{0, 6 - |\mathsf{T} - 12|\}$ is denoted as $tri(\mathsf{T})$ in Figure 5.1. We then set $\gamma_3 = 0.25\log 2$, following the same reasoning as for γ_1 and γ_2; and we choose $\sigma_2 = 1$ to control the spread of W.

5.3 Building the Model in Stan

Creating a Stan model for a BN is very simple: we just need to formalise the graphic and probabilistic representation in the previous section using the Stan language syntax. At its core, Stan is a *sampler*, which means that its key strength is the ability to draw random samples from a distribution given some information. A more in-depth technical discussion of how this is achieved can be found in Section 7.2.

For the current example, we use Stan for two tasks:

1. drawing random samples for A, I, S, O, T, W from the BN given a set of parameter values for their distributions; and

2. drawing samples for the posterior distribution of the parameters given a set of observations for A, I, S, O, T and W.

The first task entails generating data from a BN, much like the rbn function from **bnlearn**. The second task involves estimating the parameters of the local distributions from the generated data and comparing their posterior distributions with the respective values we chose in Section 5.2. In both cases we will call Stan through the **rstan** package in order to use R's facilities for plotting and summarising data.

```
library(rstan)
```

5.3.1 Generating Data

A Stan program is organised into *blocks* in which inputs, model and output are declared. Inputs are in the data block, which in this case contains the parameters for the local distributions of the variables. The outputs, that is, the random observations for the variables, are generated in the generated quantities block.[5]

```
stancode <- 'data {
  vector[2] Ap; // shape parameters for the beta distribution.
  vector[4] Ip; // probabilities for incident types.
  vector[6] Sp; // regression coefficients, logistic regression.
  vector[2] Op; // parameters for the beta distribution.
  vector[2] Tp; // parameters for the von Mises distribution.
  vector[5] Wp; // regression coefficients, log-linear regression.
}
generated quantities {
  real A;
  int I;
  real S;
  real O;
  real W;
  real T;
  A = ceil(beta_rng(Ap[1], Ap[2]) * 100);
  I = categorical_rng(Ip);
  S = 2 + binomial_rng(10, inv_logit(Sp[1] + A * Sp[2] + Sp[2 + I]));
  O = beta_rng(Op[1], Op[2]);
  T = (von_mises_rng(Tp[1], Tp[2]) + pi()) / (2 * pi()) * 24;
  W = lognormal_rng(Wp[1] + O * Wp[2] + (12 - S) * Wp[3] +
                    fmax(6 - fabs(T - 12), 0) * Wp[4], Wp[5]);
}'
```

After saving the code into a long, multi-line string, we pass it to stan_model which compiles it and makes the model accessible through R. Another option is to save the code into a file and load it with the file argument.

```
data.model <- stan_model(model_code = stancode)
```

We then pass the model object to the sampling function which, as the name suggests, performs the random sampling.

---

[5]Mathematical constants use the same notation as functions, so π is pi() and not pi.

```
params <- list(
  Ap = c(1.1, 1.5),
  Ip = c(0.075, 0.50, 0.25, 0.175),
  Sp = c(7, -0.05, 0, -4, -3, -1),
  Op = c(12, 2),
  Tp = c(0.001, 2),
  Wp = c(log(20), 2 * log(2), -0.5 * log(2), 0.25 * log(2), 1)
)
fit <- sampling(data.model, algorithm = "Fixed_param",
         data = params, thin = 25, iter = 50000, seed = 42)
nodes <- c("A", "I", "S", "O", "T", "W")
aewait <- as.data.frame(extract(fit)[nodes])
```

To make the model more general, the values of the parameters are not included in the Stan code: parameters are declared, but not initialised. Instead, we initialise them to the values we discussed in Section 5.2 using the data argument. This approach makes it easy to perform a *sensitivity analysis*, and to study the effect of different parameter values on the BN. (This is also a way of implementing approximate inference, covered in Section 6.6.2.2.) We also set the random seed with the seed argument to ensure the reproducibility of the generated samples. In addition, we make Stan aware that the parameters themselves are fixed (algorithm = "Fixed_param"); and that it should only keep 1 sample every 25 to make samples approximately independent (thin = 25, see Section 7.2.2 for details).

Finally, we transform I into a factor with self-explanatory labels to make for more idiomatic R code in the exploratory data analysis below.

```
aewait$I <- factor(aewait$I,
           labels = c("domestic", "road", "work", "other"))
head(aewait)
      A         I  S      O     T      W
1 40      work 12 0.913 11.09 182.66
2 65      road  7 0.793  7.99   7.14
3 32  domestic 12 0.954 15.20  14.75
4 31      road 10 0.751  8.71  54.23
5  4     other 12 0.918  9.90 376.35
6 48     other 11 0.770 11.45  34.06
```

5.3.2 Exploring the Variables

Several aspects of this BN are interesting to investigate, either from a theoretical or from a practical point of view. Here we will focus on checking that the parameter values we chose result in a realistic model: sanity-checking models built from expert knowledge is crucial to be able to trust any conclusions we later draw from them.

Firstly, we look at the frequency of the different brackets of trauma scores using their empirical distribution function.

```
S.cdf <- ecdf(aewait$S)
S.cdf(c(3, 7, 10, 11, 12))
[1] 0.0283 0.2050 0.4703 0.6550 1.0000
```

As we would expect, only a small fraction of patients are classified with a black code (less than 3% have $S \leqslant 3$), and few are in the lower half of the red code bracket (about 16% have $S \in [3,7]$). About 25% of patients are less-serious red codes and another 20% are yellow codes, leaving the remaining 35% as green codes. This is roughly in line with the 30% reported by the NHS some years ago outside of the peak of the flu season.

Secondly, we look into the waiting times in triage. About 89% of the patients are seen within 4 hours (that is, 240 minutes), in line with NHS figures; and about half of the patients are seen within 1 hour.

```
W.cdf <- ecdf(aewait$W)
W.cdf(c(10, 30, 60, 120, 180, 240))
[1] 0.119 0.310 0.488 0.691 0.801 0.864
```

Patients in critical conditions ($S \leqslant 3$) are seen as soon as they arrive: 73% within 10 minutes, almost all within 30 minutes.

```
nS <- length(which(aewait$S <= 3))
length(which((aewait$S <= 3) & (aewait$W < 10))) / nS
[1] 0.726
length(which((aewait$S <= 3) & (aewait$W < 30))) / nS
[1] 0.92
```

On the other hand, patients with yellow and green codes ($S \geqslant 10$) make up the vast majority of those waiting for more than 4 hours.

```
nW <- length(which(aewait$W > 240))
length(which((aewait$S >= 10) & (aewait$W > 240))) / nW
[1] 0.96
```

5.4   Estimating the Parameters in Stan

After generating random observations from the BN we constructed using the distributions and the parameters from Section 5.2, we can explore how to use Stan to do the converse: drawing random samples from the joint posterior distribution of the parameters given a set of random observations for A, I, S, O, T and W. For simplicity, we will use the data stored in the `aewaits` data frame from the previous section.

As before, we start by creating the Stan program. The relevant blocks are the data block, in which we enter the sample size (n), the variables and some transformed variables that we will precompute in R; the parameters block, which contains the parameters whose posterior distribution will be sampled; and the model block, which contains the probabilistic representation of the BN.

```
stancode <- 'data {
  int<lower=1> n;
  vector<lower=0,upper=1>[n] A;
  int<lower=1,upper=4> I[n];
  matrix[n, 4] Im;
  int<lower=0,upper=10> S[n];
  vector<lower=0,upper=10>[n] Scomp;
  vector<lower=0,upper=1>[n] O;
  vector<lower=-pi(),upper=pi()>[n] T;
  vector<lower=0,upper=6>[n] Ttri;
  vector<lower=0>[n] W;
}
parameters {
  real<lower=0> Ap[2];
  simplex[4] Ip;
  vector[4] Sbeta;
  real Sa;
  real<lower=0> Op[2];
  real<lower=-pi(),upper=pi()> Tmu;
  real<lower=0> Ts;
  real Wp[4];
  real<lower=0> Ws;
}
model {
  A ~ beta(Ap[1], Ap[2]);
  I ~ categorical(Ip);
  S ~ binomial_logit(10, A * Sa + Im * Sbeta);
  O ~ beta(Op[1], Op[2]);
  T ~ von_mises(Tmu, Ts);
  W ~ lognormal(Wp[1] + O * Wp[2] + Scomp * Wp[3] + Ttri * Wp[4], Ws);
}'
```

Since Stan has no understanding of R objects, we have to manually pass the sample size (which is not read from the data), construct the model matrix for the Incident and later convert I into an integer variable.

```
Im <- model.matrix(~ I, data = aewait)
I <- as.integer(aewait$I)
```

This matrix will contain one column for the intercept, and one 0/1 contrast for each of road, work and other; the Sbeta parameter vector in the Stan code will then contain the $\beta_0, \beta_2, \beta_3$ and β_4 parameters from Equation (5.1). (β_2 is stored separately in Sa). As for the other parameters, Tmu and Ts are the two element of Tp from the model in Section 5.3.1; and Ws is the last element of Wp. Here we declare them separately to be able to specify the respective domains in the parameters block.

For simplicity, we rescale A to $[0, 1]$ (the natural domain of the beta distribution), T to $[-\pi, \pi]$ (the natural domain of the von Mises distribution) and S to $[0, 10]$; and we precompute the complement of S and the triangular transformation of T. We also take care that no values of A lie on the boundary of the domain: this, as well as the presence of any missing data, would prevent sampling from initialising the MCMC sampler.

```
A <- as.numeric(aewait$A / 100)
A[A == 0] <- .Machine$double.eps
A[A == 1] <- 1 - .Machine$double.eps
S <- as.integer(aewait$S - 2)
Scomp <- 10 - S;
T <- aewait$T / 24 * (2 * pi) -pi
Ttri <- pmax(6 - abs(aewait$T - 12), 0)
```

All this preprocessing could be done instead in Stan in the generated data and generated parameters code blocks, but it is faster and more convenient to do it in R.

After these preparations, we compile the model and we sample from parameter values from their posterior distributions. The number of iterations

label	true val.	mean $\pm$ st.dev.	label	true val.	mean $\pm$ st.dev.
Ap[1]	1.1	1.078 ± 0.022	Sp[6]	-1	-0.986 ± 0.174
Ap[2]	1.5	1.407 ± 0.029	Op[1]	12	11.71 ± 0.263
Ip[1]	0.075	0.075 ± 0.004	Op[2]	2	1.937 ± 0.039
Ip[2]	0.5	0.492 ± 0.008	Tp[1]	0.001	-0.017 ± 0.013
Ip[3]	0.25	0.257 ± 0.007	Tp[2]	2	1.983 ± 0.038
Ip[4]	0.175	0.176 ± 0.006	Wp[1]	2.996	3.251 ± 0.009
Sp[1]	7	7.035 ± 0.166	Wp[2]	1.386	1.108 ± 0.009
Sp[2]	-0.05	-0.049 ± 0.001	Wp[3]	-0.347	-0.347 ± 0.009
Sp[3]	0	0 ± 0	Wp[4]	0.173	0.174 ± 0.009
Sp[4]	-4	-4.092 ± 0.161	Wp[5]	1	1.015 ± 0.012
Sp[5]	-3	-3.11 ± 0.174			

Table 5.1
True values and posterior means $\pm$ standard deviations for estimates of the parameters of the BN, labelled with the names they had in the Stan code in Section 5.3.1.

is smaller than before (3500 vs 50000) because this Stan model is sampling from the joint distribution of 20 parameters instead of that of six variables; hence it is noticeably slower.

```
parameters.model <- stan_model(model_code = stancode)
fit <- sampling(model, iter = 3500, seed = 42, thin = 25,
        data = list(n = nrow(aewait), A = A, I = I, S = S,
                    Scomp = Scomp, O = aewait$O, T = T,
                    Ttri = Ttri, W = aewait$W))
aeparams <- as.data.frame(extract(fit))
```

The `aeparams` data frame is too large to display in print, so we summarise it in Table 5.1 with the posterior mean and the standard deviation of each parameter. All the posterior means are within 1–2 standard deviations of the original values from Section 5.2, suggesting that MCMC sampling was successful in approximating the posterior distribution. This level of agreement was achieved without specifying any prior distribution for the parameters; we just let Stan use non-informative priors.

In a real-world analysis, we would have used the information we constructed the BN from to define informative priors in the `model` block. For instance, we could have said

```
Op[1] ~ exponential(0.1);
Op[2] ~ exponential(0.5);
```

since both `Op` parameters are positive. The prior expected value of `O` would be 0.9, and the expected values of `Op[1]` and `Op[2]` would result in a density with a shape similar to that in the middle panel of Figure 5.2. Giving Stan such information about the distribution of the parameters in this way can potentially make for faster sampling and more accurate estimates.

5.5   Further Reading

Many books have been published in recent years on the topic of Stan and Bayesian inference: two notable examples are "Statistical Rethinking" (McElreath, 2016) and "Doing Bayesian Data Analysis" (Kruschke, 2014).

Exercises

Exercise 5.1 *Consider again the* ergoStool *data from Exercise 3.3, which is originally from Pinheiro and Bates (2000) and which describes a randomised-block experiment which recorded the effort (E) required by each of nine different subjects (S) to arise from each of four types of stools (T).*

> *1. Load the* ergoStool *data for this experiment from the* **nlme** *package, and rename the variables to E, S, T.*

> *2. Create a BN in Stan in which S, T are multinomials and E is a normal random variable, and use the* ergoStool *data to estimate their parameters and sample from their posterior distributions. The structure of the BN should be $S \rightarrow E \leftarrow T$.*

> *3. Summarise the posterior samples with their mean and standard deviation, and comment on the precision of the posterior estimates.*

Exercise 5.2 *Consider a bus stop with two bus lines B1 and B2, and a tram line T. Buses on these lines have a number of passengers that can be modelled as $Pois(\lambda_1)$ and $Pois(\lambda_2)$, respectively; passengers may decide to get off with probabilities p_1 and p_2. Trams will have $Pois(\lambda_3)$ passengers, which may decide to get off with probability p_3. Passengers disembarked from B1 and B2 may then decide to board T with probabilities q_1 and q_2.*

Create a Stan model that can be used to simulate how many passengers will be on the tram when it departs, assuming $\lambda_1 = \lambda_2 = 25$, $\lambda_3 = 20$, $p_1 = 0.6$, $p_2 = 0.3$, $p_3 = 0.1$, $q_1 = q_2 = 0.15$.

Exercise 5.3 *Consider again the bus and tram lines from Exercise 5.2. Construct a Stan model that simulates from the posterior distribution of the parameters assuming normal priors for λ_1, λ_2, λ_3 and beta priors for p_1, p_2, p_3, q_1 and q_2. Use the data simulated in Exercise 5.2 to run the model.*

6

Theory and Algorithms for Bayesian Networks

In this chapter we will provide the theoretical foundations underpinning the classes of BNs we explored in the previous chapters. In particular, we will introduce the formal definition of a BN and its fundamental properties. We will then show how these properties provide a rigorous foundation for BN learning and inference.

6.1 Conditional Independence and Graphical Separation

BNs are a class of *graphical models*, which allow an intuitive representation of the probabilistic structure of multivariate data using graphs. We introduced them in Chapter 1 as the combination of:

- A set of random variables $\mathbf{X} = \{X_1, X_2, \ldots, X_p\}$ describing the quantities of interest. The multivariate probability distribution of $\mathbf{X}$ is called the *global distribution* of the data, while the univariate distributions associated with the $X_i \in \mathbf{X}$ are called *local distributions*.

- A *directed acyclic graph* (DAG), denoted $G = (\mathbf{V}, A)$. Each node $v \in \mathbf{V}$ is associated with one variable X_i. The directed arcs $a \in A$ that connect them represent direct probabilistic dependencies; so if there is no arc connecting two nodes the corresponding variables are either independent or conditionally independent given a subset of the remaining variables.

The link between the graphical separation (denoted $\perp\!\!\!\perp_G$) induced by the absence of a particular arc and probabilistic independence (denoted $\perp\!\!\!\perp_P$) provides a direct and easily interpretable way to express the relationships between the variables. Following the seminal work of Pearl (1988), we distinguish three possible ways in which the former maps to the latter.

Definition 6.1 (Maps) *Let M be the dependence structure of the probability distribution P of $\mathbf{X}$, that is, the set of conditional independence relationships linking any triplet $\mathbf{A}$, $\mathbf{B}$, $\mathbf{C}$ of subsets of $\mathbf{X}$. A graph G is a dependency map (or D-map) of M if there is a one-to-one correspondence* *between the random variables in* $\mathbf{X}$ *and the nodes* $\mathbf{V}$ *of* G *such that for all disjoint subsets* $\mathbf{A}$, $\mathbf{B}$, $\mathbf{C}$ *of* $\mathbf{X}$ *we have*

$$\mathbf{A} \perp\!\!\!\perp_P \mathbf{B} \mid \mathbf{C} \Longrightarrow \mathbf{A} \perp\!\!\!\perp_G \mathbf{B} \mid \mathbf{C}. \tag{6.1}$$

Similarly, G *is an independency map (or I-map) of* M *if*

$$\mathbf{A} \perp\!\!\!\perp_P \mathbf{B} \mid \mathbf{C} \Longleftarrow \mathbf{A} \perp\!\!\!\perp_G \mathbf{B} \mid \mathbf{C}. \tag{6.2}$$

G *is said to be a perfect map of* M *if it is both a D-map and an I-map, that is*

$$\mathbf{A} \perp\!\!\!\perp_P \mathbf{B} \mid \mathbf{C} \Longleftrightarrow \mathbf{A} \perp\!\!\!\perp_G \mathbf{B} \mid \mathbf{C}, \tag{6.3}$$

and in this case G *is said to be faithful or isomorphic to* M.

In the case of a D-map, the probability distribution of $\mathbf{X}$ determines which arcs are present in the DAG G. Nodes that are connected (that is, not separated) in G correspond to dependent variables in $\mathbf{X}$. However, nodes that are separated in G do not necessarily correspond to conditionally independent variables in $\mathbf{X}$. On the other hand, in the case of an I-map we have that the arcs present in G determine which variables are conditionally independent in $\mathbf{X}$. Therefore, nodes that are found to be separated in G correspond to conditionally independent variables in $\mathbf{X}$, but nodes that are connected in G do not necessarily correspond to dependent variables in $\mathbf{X}$. In the case of a perfect map, there is a one-to-one correspondence between graphical separation in G and conditional independence in $\mathbf{X}$.

The graphical separation in Definition 6.1 is established using *d-separation*, which we first introduced in Section 1.6.1. It is formally defined as follows.

Definition 6.2 (d-separation) *If* $\mathbf{A}$, $\mathbf{B}$ *and* $\mathbf{C}$ *are three disjoint subsets of nodes in a DAG* G, *then* $\mathbf{C}$ *is said to d-separate* $\mathbf{A}$ *from* $\mathbf{B}$, *denoted* $\mathbf{A} \perp\!\!\!\perp_G \mathbf{B} \mid \mathbf{C}$, *if along every path between a node in* $\mathbf{A}$ *and a node in* $\mathbf{B}$ *there is a node* v *satisfying one of the following two conditions:*

> *1. v has converging arcs (i.e., there are two arcs pointing to v from the adjacent nodes in the path) and neither v nor any of its descendants (i.e., the nodes that can be reached from v) are in $\mathbf{C}$.*

> *2. v is in $\mathbf{C}$ and does not have converging arcs.*

As an example, consider again the three fundamental connections shown in Figure 1.3, Section 1.6.1. The first was a serial connection from Sex to Education to Residence ($\mathsf{S} \to \mathsf{E} \to \mathsf{R}$), and we were investigating $\mathsf{S} \perp\!\!\!\perp_G \mathsf{R} \mid \mathsf{E}$. The node E, which plays the role of $v \in \mathbf{C}$ in Definition 6.2, satisfies the second condition and d-separates S and R. As a result, we can conclude that $\mathsf{S} \perp\!\!\!\perp_G \mathsf{R} \mid \mathsf{E}$ holds and, in turn, we can determine that S and R are conditionally independent ($\mathsf{S} \perp\!\!\!\perp_P \mathsf{R} \mid \mathsf{E}$) using Equation (6.2) in Definition 6.1. Similarly, we can conclude that $\mathsf{O} \perp\!\!\!\perp_G \mathsf{R} \mid \mathsf{E}$ and $\mathsf{O} \perp\!\!\!\perp_P \mathsf{R} \mid \mathsf{E}$ holds for the divergent connection formed by Education, Occupation and Residence $(\mathsf{O} \leftarrow \mathsf{E} \to \mathsf{R})$. On the other hand, in the convergent connection formed by Age, Sex and Education $(\mathsf{A} \to \mathsf{E} \leftarrow \mathsf{S})$ we have that $\mathsf{A} \not\!\perp_G \mathsf{S} \mid \mathsf{E}$. Unlike the serial and divergent connections, the node in the middle of the connection does not d-separate the other two since E does not satisfy any of the two conditions in Definition 6.2.

6.2 Bayesian Networks

Having defined a criterion to determine whether two nodes are connected or not, and how to map those connections (or the lack thereof) to the probability distribution of $\mathbf{X}$, we are now ready to formally define BNs.

Definition 6.3 (BNs) *Given a probability distribution P on a set of variables $\mathbf{X}$, a DAG $G = (\mathbf{X}, A)$ is called a BN and denoted $\mathcal{B} = (G, \mathbf{X})$ if and only if G is a minimal I-map of P, so that none of its arcs can be removed without destroying its I-mapness.*

This definition highlights two fundamental properties of BNs. First, assuming that G is an I-map leads to the general formulation of the decomposition of the global distribution $\Pr(\mathbf{X})$ introduced in Equation (1.1) on page 7:

$$\Pr(\mathbf{X}) = \prod_{i=1}^{p} \Pr(X_i \mid \Pi_{X_i}), \qquad (6.4)$$

where Π_{X_i} is the set of the parents of X_i. If X_i has two or more parents it depends on their joint distribution, because each pair of parents forms a convergent connection centred on X_i and we cannot establish their independence. This decomposition is preferable to that obtained from the *chain rule*,

$$\Pr(\mathbf{X}) = \prod_{i=1}^{p} \Pr(X_i \mid X_{i+1}, \ldots, X_p) \qquad (6.5)$$

because the conditioning sets are typically smaller. Only when the ordering of the variables is topological the chain rule simplifies to the decomposition in Equation (6.4). In the general case it is more difficult to write, even for the simple discrete and Gaussian BNs presented in Chapters 1 and 2.

Another result along the same lines is called the *local Markov property*, which can be combined with the chain rule above to get the decomposition in Equation (6.4).

Definition 6.4 (Local Markov property) *Each node X_i is conditionally independent of its non-descendants (e.g., nodes X_j for which there is no path from X_i to X_j) given its parents.*

Compared to the decomposition above, the local Markov property highlights the fact that parents are not completely independent from their children in the BN: a trivial application of Bayes' theorem to invert the direction of the conditioning shows how information on a child can change the distribution of the parent.

Second, assuming that G is an I-map also means that serial and divergent connections result in equivalent factorisations of the variables involved. It is easy to show that

$$\underbrace{\Pr(X_i)\Pr(X_j \mid X_i)\Pr(X_k \mid X_j)}_{\text{serial connection}} = \Pr(X_j, X_i)\Pr(X_k \mid X_j) =$$

$$= \underbrace{\Pr(X_i \mid X_j)\Pr(X_j)\Pr(X_k \mid X_j)}_{\text{divergent connection}}. \quad (6.6)$$

Then $X_i \to X_j \to X_k$ and $X_i \gets X_j \to X_k$ are equivalent. As a result, we can have BNs with different arc sets that encode the same conditional independence relationships and represent the same global distribution in different (but probabilistically equivalent) ways. Such DAGs are said to belong to the same *equivalence class*.

Theorem 6.1 (Equivalence classes) *Two DAGs defined over the same set of variables are equivalent if and only if they have the same skeleton (that is, the same underlying undirected graph) and the same v-structures.*

In other words, the only arcs whose directions are important are those that are part of one or more *v-structures*.

Definition 6.5 (V-structures) *A convergent connection $X_i \to X_k \gets X_j$ is called a v-structure if there is no arc connecting X_i and X_j. In addition, X_k is often called the collider node and the connection is then called an unshielded collider, as opposed to a shielded collider in which either $X_i \to X_j$ or $X_i \gets X_j$.*

As underlined both in Section 1.6.1 and in Section 6.1, convergent connections that are also v-structures have different characteristics than serial and divergent connections for both graphical separation and probabilistic independence. Therefore, the directions of their arcs cannot be changed without altering the global distribution.

Consider, for example, the graphs in Figure 6.1. Using **bnlearn**, we can create the DAG in the top left panel,

```
X <- paste0("[X1][X3][X5][X6|X8][X2|X1][X7|X5][X4|X1:X2][X8|X3:X7]",
            "[X9|X2:X7][X10|X1:X9]")
dag <- model2network(X)
```

and we can get its skeleton (top-right panel) and v-structures as follows.

```
skel <- skeleton(dag)
```

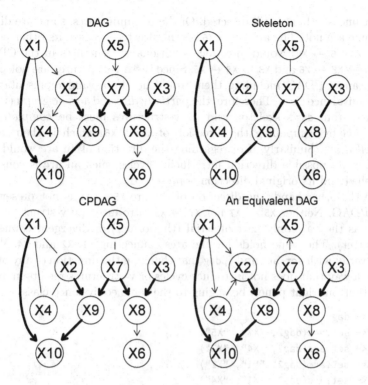

Figure 6.1
A DAG (top left), its underlying undirected graph (the *skeleton*, top right),
its CPDAG (bottom left) and another DAG in the same equivalence class
(bottom right). Arcs that belong to a v-structure are drawn with a thicker
line width.

```
vstructs(dag)
       X    Z     Y
[1,] "X1" "X10" "X9"
[2,] "X3" "X8"  "X7"
[3,] "X2" "X9"  "X7"
```

These two quantities identify the equivalence class dag belongs to, which
is represented by the *completed partially directed acyclic graph* (CPDAG)
shown in the bottom-left panel. We can obtain it from dag with the cpdag
function.

```
cp1 <- cpdag(dag)
```

Following Theorem 6.1, not all arcs in this graph are directed: therefore, it is
not a DAG because it is only partially directed.

Arcs that were part of one or more v-structures in dag, highlighted with a thicker line width, are still directed. Of the remaining arcs, some are directed and some are not. The arc X8 → X6 is still directed because its other possible direction (X6 → X8) would introduce additional v-structures in the CPDAG, e.g., X7 → X8 ← X6 and X3 → X8 ← X6. Since these v-structures are not present in dag, any CPDAG including them would not be a valid representation of dag's equivalence class. Therefore, the *partially directed acyclic graph* (PDAG) obtained from dag's skeleton and its v-structures must be *completed* into a CPDAG by *compelling* the direction of X8 → X6, which is then called a *compelled arc*. Similarly, if disregarding the direction of an arc would result in one of its possible direction introducing cycles, such an arc is considered compelled and its original direction is preserved.

On the other hand, the direction of the arc X5 → X7 is not preserved in the CPDAG. Neither X5 → X7 nor X7 → X5 introduce any v-structure in the graph, as they are only part of serial (the former) or divergent connections (the latter). The same holds for the arcs connecting X1, X2 and X4. We can easily verify with set.arc and cpdag that changing the direction of any of those arcs in a way that does not introduce cycles or v-structures results in another DAG (bottom right panel) belonging to the same equivalence class.

```
dag2 <- dag
dag2 <- set.arc(dag2, "X7", "X5")
dag2 <- set.arc(dag2, "X4", "X2")
dag2 <- set.arc(dag2, "X1", "X2")
dag2 <- set.arc(dag2, "X1", "X4")
cp2 <- cpdag(dag2)
all.equal(cp1, cp2)
[1] TRUE
```

It is important to note that even though X1 → X4 ← X2 is a convergent connection, it is not a v-structure because X1 and X2 are connected by X1 → X2. As a result, we are no longer able to identify which nodes are the parents in the connection. For example, we can show that $\{X1, X2, X4\}$ have exactly the same probability distribution in the two DAGs in Figure 6.1:

$$\underbrace{\Pr(X1)\Pr(X2 \mid X1)\Pr(X4 \mid X1, X2)}_{X1 \to X4 \leftarrow X2, \, X1 \to X2} = \Pr(X1)\frac{\Pr(X2, X1)}{\Pr(X1)}\frac{\Pr(X4, X1, X2)}{\Pr(X1, X2)} =$$

$$= \Pr(X1)\Pr(X4, X2 \mid X1) = \underbrace{\Pr(X1)\Pr(X2 \mid X4, X1)\Pr(X4 \mid X1)}_{X4 \to X2 \leftarrow X1, \, X1 \to X4}. \quad (6.7)$$

Therefore, the fact that the two parents in a convergent connection are not connected by an arc is crucial in the identification of the correct CPDAG.

6.3 Markov Blankets

The decomposition of the global distribution in Equation (6.4) provides a convenient way to split $\mathbf{X}$ into manageable parts, and identifies in the parents of each node the set of conditioning variables of each local distribution. This is indeed very useful for manually constructing a BN or for learning it from data. However, for the purpose of performing inference it may be preferable to use Bayes' theorem as suggested by Definition 6.4 and to also consider the children of a node to increase the amount of information extracted from the BN.

Going back to the first DAG in Figure 6.1, if we want to perform a query on X9 instead of using

$$X9 \sim \Pr(X9 \mid X2, X7) \tag{6.8}$$

which incorporates only the information provided by X2 and X7, we may prefer to include more nodes to use the information encoded in the BN to a greater extent, and make inference more powerful as a consequence.

In the limit case, we could decide to condition on all the other nodes in the BN,

$$X9 \sim \Pr(X9 \mid X1, X2, X3, X4, X5, X6, X7, X8, X10), \tag{6.9}$$

to prevent any bias or loss of power. However, this is unfeasible for most real-world problems, and in fact it would not be much different from using the global distribution. Instead, we can use d-separation to reduce the set of conditioning nodes we need to consider. For example, adding X5 to the conditioning variables in the example Equation (6.8) is pointless, because it is d-separated (and therefore conditionally independent) from X9 by $\{X2, X7, X10\}$:

```
dsep(dag, x = "X9", y = "X5", z = c("X2", "X7", "X10"))
[1] TRUE
```

The minimal subset of nodes we should condition on is called the *Markov blanket*, and is defined as follows.

Definition 6.6 (Markov blanket) *The Markov blanket of a node $A \in \mathbf{V}$ is the minimal subset $\mathbf{S}$ of $\mathbf{V}$ such that*

$$A \perp\!\!\!\perp_G \mathbf{V} - \mathbf{S} - A \mid \mathbf{S}. \tag{6.10}$$

Corollary 6.1 (Markov blankets and faithfulness) *Assuming faithfulness, Definition 6.6 implies $\mathbf{S}$ is the minimal subset of $\mathbf{V}$ such that*

$$A \perp\!\!\!\perp_P \mathbf{V} - \mathbf{S} - A \mid \mathbf{S}. \tag{6.11}$$

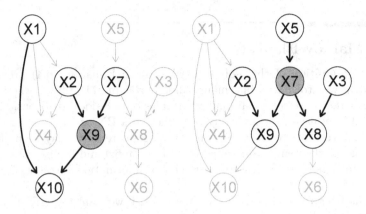

Figure 6.2
The Markov blankets of nodes X9 (on the left) and X7 (on the right).

Corollary 6.1 provides a formal definition of the set we are looking for: it is the subset of nodes that makes the rest redundant when performing inference on a given node. In other words, it casts Markov blanket identification as a general *feature selection* or *variable selection* problem. Assuming faithfulness makes sure that the set is indeed minimal. Assuming the BN is just an I-map, as in the standard definition, does not guarantee that all the nodes in the Markov blanket are really required to get complete probabilistic independence because not all nodes that are shown to be connected in G are in fact probabilistically dependent.

Theorem 6.2 (Composition of the Markov blanket) *The Markov blanket of a node A is the set consisting of the parents of A, the children of A and all the other nodes sharing a child with A.*

Theorem 6.2 identifies which nodes should be included in the Markov blanket to d-separate the target node from the rest of the DAG. Parents and children are required to block the paths that satisfy the first condition in Definition 6.2. Nodes that share a child with the target node, sometimes called *spouses*, are required to block the paths that satisfy the second condition. The target node itself is not part of its Markov blanket.

As an example, the Markov blankets of nodes X9 and X7 are shown in Figure 6.2. We can identify the nodes that belong to each with the mb function from **bnlearn**.

```
mb(dag, node = "X9")
[1] "X1"  "X10" "X2"  "X7"
mb(dag, node = "X7")
[1] "X2" "X3" "X5" "X8" "X9"
```

Using the functions `parents` and `children` we can also check that the Markov blanket of X9 is composed by the nodes identified by Definition 6.6. First, we identify the parents and the children of X9.

```
par.X9 <- parents(dag, node = "X9")
ch.X9 <- children(dag, node = "X9")
```

Then we identify all the parents of the children of X9 by calling the `parents` function for each child through `sapply`.

```
sp.X9 <- sapply(ch.X9, parents, x = dag)
```

We must remove X9 itself from `sp.X9`, because a node by definition is not part of its own Markov blanket. Finally, we merge parX9, `ch.X9` and `sp.X9` to form the Markov blanket.

```
sp.X9 <- setdiff(sp.X9, "X9")
union(union(par.X9, ch.X9), sp.X9)
 [1] "X2"  "X7"  "X10"  "X1"
```

Equivalently, we might have used `nbr` to identify both parents and children in a single function call, as they are the neighbours of X9. The Markov blanket of X7 can be similarly investigated.

We can also test whether the Markov blanket of X9 d-separates X9 from all the other nodes in `dag`.

```
V <- setdiff(nodes(dag), "X9")
S <- mb(dag, "X9")
sapply(setdiff(V, S), dsep, bn = dag, y = "X9", z = S)
   X3    X4    X5    X6    X8
 TRUE TRUE TRUE TRUE TRUE
```

Similarly, for X7 we have

```
V <- setdiff(nodes(dag), "X7")
S <- mb(dag, "X7")
sapply(setdiff(V, S), dsep, bn = dag, y = "X7", z = S)
   X1   X10    X4    X6
 TRUE TRUE TRUE TRUE
```

as expected.

Furthermore, it can be shown as a corollary of Theorem 6.2 that Markov blankets are symmetric.

Corollary 6.2 (Symmetry of Markov blankets) *Theorem 6.2 defines a symmetric relationship: if node A is in the Markov blanket of B, then B is in the Markov blanket of A.*

So, for example, if we take each node in the Markov blanket S of X7 in turn and identify its Markov blanket, we have that X7 belongs to each of those Markov blankets.

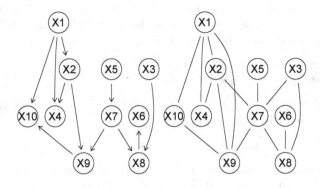

Figure 6.3
The DAG from Figures 6.1 and 6.2 (left) and its moral graph (right). Some
nodes have been moved to give a clearer view of the relationships.

```
belongs <- logical(0)
for (node in S)
  belongs[node] <- "X7" %in% mb(dag, node)
belongs
    X2   X3   X5   X8   X9
  TRUE TRUE TRUE TRUE TRUE
```

6.4 Moral Graphs

In Section 6.2 we introduced an alternative graphical representation of the
DAG underlying a BN: the CPDAG of the equivalence class the BN belongs
to. Another graphical representation that can be derived from the DAG is the
moral graph.

The moral graph is an undirected graph that is constructed as follows:

1. connecting the parents in each v-structure with an undirected arc;

2. removing the direction of all other arcs, effectively replacing them
   with undirected arcs.

This transformation is called *moralisation* because it "marries" non-adjacent
parents sharing a common child. The moral graph of our example dag is shown
in Figure 6.3; we created it with the moral function as follows.

```
mg1 <- moral(dag)
```

It is important to note that different DAGs may result in identical moral graphs because of the nature of the transformations above. For instance, adding an arc from X7 to X3 to dag does not alter its moral graph.

```
all.equal(moral(dag),
          moral(set.arc(dag, from = "X7", to = "X3")))
 [1] TRUE
```

It is also easy, for the sake of the example, to perform moralisation manually. First, we identify the v-structures in the DAG and we link the parents within each v-structure with an undirected arc (that is, an edge).

```
mg2 <- dag
vs <- vstructs(dag)
for (i in seq(nrow(vs)))
  mg2 <- set.edge(mg2, from = vs[i, "X"], to = vs[i, "Y"],
            check.cycles = FALSE)
```

This step appears to introduce potential cycles in the resulting PDAG. However, we can safely ignore them since we are going to construct the undirected graph underlying mg2, thus replacing each directed arc with an undirected one.

```
mg2 <- skeleton(mg2)
all.equal(mg1, mg2)
 [1] TRUE
```

Moralisation has several uses. Firstly, it provides a simple way to transform a BN into the corresponding *Markov network*, a graphical model using undirected graphs instead of DAGs to represent dependencies. These models have a long history in statistics, and have been studied in classic monographs such as Whittaker (1990) and Lauritzen (1996). On the one hand, in a Markov network all dependencies are explicitly represented, even those that would be implicitly implied by v-structures in a BN. On the other hand, the addition of undirected arcs makes the graph structure less informative, because we cannot tell that nodes that were parents in a v-structure are marginally independent when we are not conditioning on their common child. Secondly, moral graphs provide the foundation for exact inference on BNs through the junction tree algorithm, which will be introduced in Section 6.6.2.

Thirdly, moral graphs provide a way of programmatically checking whether d-separation holds for some nodes in a few, efficient steps:

1. keep the nodes we are checking and those in the d-separating set, and remove all the nodes that are not ancestors (*i.e.*, parents, parents' parents, etc.) of any nodes in these two sets;

2. construct the moral graph of the resulting subgraph;

3. check whether there is an open (undirected) path that is not blocked (that is, that does not pass through) the nodes in the d-separating set.

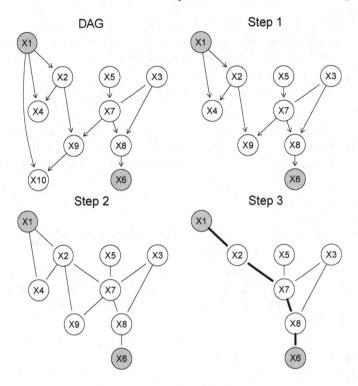

Figure 6.4
Graphical representation of the steps required to show that X1 and X6 are d-separated by X4 and X9.

As a practical example, consider once more the DAG in Figure 6.1 (now Figure 6.4, top-left panel): we would like to check whether X1 and X6 are d-separated by X4 and X9.

```
dsep(dag, x = "X1", y = "X6", z = c("X4", "X9"))
 [1] FALSE
```

In the first step (Figure 6.4, top-right panel), we identify the ancestors of X1, X6, and those of X4 and X9.

```
an.nodes = union(ancestors(dag, "X1"), ancestors(dag, "X6"))
an.nodes
 [1] "X8" "X3" "X7" "X5"
an.sepset = union(ancestors(dag, "X4"), ancestors(dag, "X9"))
an.sepset
 [1] "X2" "X1" "X7" "X5"
```

Then we take an.nodes and an.sepset and we construct a subgraph of dag that contains only these nodes along with the nodes being checked and those in the d-separating set, removing the rest.

```
sub.nodes = union(c("X1", "X6"), an.nodes)
sub.sep = union(c("X4", "X9"), an.sepset)
sub.dag = subgraph(dag, union(sub.nodes, sub.sep))
```

In the second step we construct the moral graph of this subgraph (Figure 6.4, bottom-left panel),

```
sub.moral = moral(sub.dag)
```

and finally, in the last step, we remove the d-separating nodes (to make sure we do not pass through them) and check whether there is a path connecting X1 and X4.

```
final.nodes = setdiff(nodes(sub.moral), c("X4", "X9"))
sub.final = subgraph(sub.moral, final.nodes)
path.exists(sub.final, "X1", "X6")
[1] TRUE
```

Since there is a path connecting X1 and X6 in sub.final, which is shown in bold in the bottom-right panel of Figure 6.4, we conclude that X1 and X6 are not d-separated by X4 and X9 in the original dag; in agreement with the value returned by the dsep function above.

6.5 Bayesian Network Learning

In the context of BNs, model selection and estimation are collectively known as *learning*, a name borrowed from artificial intelligence and machine learning. BN learning is usually performed as a two-step process:

1. *structure learning*, learning the structure of the DAG;

2. *parameter learning*, learning the local distributions implied by the structure of the DAG learned in the previous step.

Both steps can be performed either using the information provided by a data set or by interviewing experts in the fields relevant for the phenomenon being modelled. Combining both approaches is common. Often the prior information available on the phenomenon is not enough for an expert to completely specify a BN. Even specifying the DAG structure is often impossible, especially when a large number of variables are involved. This is the case, for example, for most applications in genetics and systems biology (because of how many components are involved in biological processes) and in the social sciences (because of lack of agreement between experts and of solid experimental evidence).

This workflow is inherently Bayesian. Consider a data set $\mathcal{D}$ and a BN $\mathcal{B} = (G, \mathbf{X})$. If we denote the parameters of the global distribution of $\mathbf{X}$ with Θ, we can assume without loss of generality that Θ uniquely identifies $\mathbf{X}$ in the parametric family of distributions chosen for modelling $\mathcal{D}$ and write $\mathcal{B} = (G, \Theta)$. BN learning can then be formalised as

$$\underbrace{\Pr(\mathcal{B} \mid \mathcal{D}) = \Pr(G, \Theta \mid \mathcal{D})}_{\text{learning}} = \underbrace{\Pr(G \mid \mathcal{D})}_{\text{structure learning}} \cdot \underbrace{\Pr(\Theta \mid G, \mathcal{D})}_{\text{parameter learning}}.$$

(6.12)

The decomposition of $\Pr(G, \Theta \mid \mathcal{D})$ reflects the two steps described above, and underlies the logic of the learning process.

Structure learning can be done in practice by finding the DAG G that maximises

$$\Pr(G \mid \mathcal{D}) \propto \Pr(G) \Pr(\mathcal{D} \mid G) = \Pr(G) \int \Pr(\mathcal{D} \mid G, \Theta) \Pr(\Theta \mid G) \, d\Theta, \quad (6.13)$$

using Bayes' theorem to decompose the posterior probability of the DAG (*i.e.*, $\Pr(G \mid \mathcal{D})$) into the product of the prior distribution over the possible DAGs (*i.e.*, $\Pr(G)$) and the probability of the data (*i.e.*, $\Pr(\mathcal{D} \mid G)$). Clearly, it is not possible to compute the latter without estimating the parameters Θ in the process; therefore, Θ has to be integrated out of Equation (6.13) to make $\Pr(G \mid \mathcal{D})$ independent of any specific choice of Θ.

The prior distribution $\Pr(G)$ provides an ideal way to introduce any prior information available on the conditional independence relationships between the variables in $\mathbf{X}$. We may, for example, require that one or more arcs should be present in or absent from the DAG, to account for the insights gained in previous studies. We may also require that some arcs, if present in the DAG, must be oriented in a specific direction when that direction is the only one that makes sense in the light of the logic underlying the phenomenon being modelled. This was the case in the examples presented in Chapters 1 and 2. In Chapter 1, Age (A) and Sex (S) cannot be the head of any arcs because they are demographic indicators. It also makes little sense to allow either A → S or S → A to be present in the DAG. Similar considerations can be made for the genetic potential (G) and the environmental potential (E) in Chapter 2. As an alternative, we can also assign probabilities of inclusion to certain arcs in certain directions (*e.g.*, $\Pr(A \to S) = 0.01$ and $\Pr(S \to A) = 0.01$, which implies that there is no arc between A and S with probability 0.98).

There are, however, limits to these approaches: specifying an informative prior over the space of DAGs (sometimes known as *structural prior*) becomes more and more difficult as the number of nodes increases. The number of possible DAGs increases super-exponentially in the number of nodes. In a DAG with p nodes we have $\frac{1}{2}p(p-1)$ possible arcs, given by the pairs of different nodes in $\mathbf{V}$. Even disregarding the arcs' directions, this means that there are $O(2^{p^2})$ possible DAGs. As a result, in most cases we tend to use simpler *reference priors* that only contain some general information about what type of DAGs we expect to learn. The simplest is the *uniform prior*,

$$\Pr(G) \propto 1 \qquad \text{for all DAGs } G,$$

which is convenient for both computational and algebraic reasons because it vanishes when maximising $\Pr(G \mid \mathcal{D})$. However, it has the undesirable property of favouring dense graphs: since all DAGs are equally probable, all states of each possible arc are equally probable as well. This means that for each pair of nodes (X_i, X_j) we have that $\Pr(X_i \to X_j) + \Pr(X_j \to X_i) = \frac{1}{3} + \frac{1}{3} = \frac{2}{3}$, and that in turn we expect a DAG to have $\frac{2}{3} \cdot \frac{1}{2}p(p-1)$ arcs.

An equally simple prior that does not suffer from this problem is the *marginal uniform* prior from Scutari (2016) that for each arc specifies

$$\Pr(X_i \to X_j) = \Pr(X_j \to X_i) \quad \text{and} \quad \Pr(X_i \to X_j) + \Pr(X_j \to X_i) = c$$

with $c \in [0, 1]$. Hence, both arc directions are equally probable, but the probability that two nodes are linked by an arc can be controlled by specifying c. The uniform prior leads to $c = \frac{2}{3}$; a "neutral" prior that makes arc presence and absence equally probable leads to $c = \frac{1}{2}$; a prior that favours sparse graphs leads to $c = \frac{1}{p-1}$ so that we expect a DAG to have $2p$ arcs. Other possible priors along these lines such as the *variable selection* prior that gives decreasing probabilities to larger parent sets are reviewed in Eggeling et al. (2019).

Computing $\Pr(\mathcal{D} \mid G)$ is also problematic from both a computational and a mathematical point of view. Starting from the decomposition into local distributions, and assuming the Θ_{X_i} are mutually independent, we can further factorise $\Pr(\mathcal{D} \mid G)$ in a similar way:

$$\Pr(\mathcal{D} \mid G) = \int \prod_{i=1}^{p} [\Pr(X_i \mid \Pi_{X_i}, \Theta_{X_i}) \Pr(\Theta_{X_i} \mid \Pi_{X_i})] \, d\Theta =$$

$$= \prod_{i=1}^{p} \left[\int \Pr(X_i \mid \Pi_{X_i}, \Theta_{X_i}) \Pr(\Theta_{X_i} \mid \Pi_{X_i}) \, d\Theta_{X_i} \right] =$$

$$= \prod_{i=1}^{p} \mathbb{E}_{\Theta_{X_i}} [\Pr(X_i \mid \Pi_{X_i})]. \quad (6.14)$$

Functions that can be factorised in this way are called *decomposable*. If all expectations can be computed in closed form, $\Pr(\mathcal{D} \mid G)$ can be computed in a reasonable time even for large data sets. This is possible for both the multinomial distribution assumed for discrete BNs (via its conjugate Dirichlet posterior) and the multivariate Gaussian distribution assumed for continuous BNs (via its conjugate Inverse Wishart distribution). For discrete BNs, $\Pr(\mathcal{D} \mid G)$ can be estimated with the *Bayesian Dirichlet equivalent uniform* (BDeu) score from Heckerman et al. (1995). Since it is the only member of the BDe family of scores in common use, it is often referred to simply as BDe. As we have seen in Chapter 1, BDe assumes a flat prior over the parameter space of each node:

$$\Pr(\Theta_{X_i} \mid \Pi_{X_i}) = \alpha_{ijk} = \frac{\alpha}{|\Theta_{X_i}|}.$$

The only parameter of BDe is the *imaginary sample size* α associated with the Dirichlet prior, which determines how much weight is assigned to the prior as the size of an imaginary sample supporting it.

Under these assumptions, BDe takes the form

$$\mathrm{BDe}(G, \mathcal{D}) = \prod_{i=1}^{p} \mathrm{BDe}(X_i, \Pi_{X_i}) =$$

$$= \prod_{i=1}^{p} \prod_{j=1}^{q_i} \left[\frac{\Gamma(\alpha_{ij})}{\Gamma(\alpha_{ij} + n_{ij})} \prod_{k=1}^{r_i} \frac{\Gamma(\alpha_{ijk} + n_{ijk})}{\Gamma(\alpha_{ijk})} \right], \quad (6.15)$$

where:

- p is the number of nodes in G;

- r_i is the number of categories for the node X_i;

- q_i is the number of configurations of the categories of the parents of X_i;

- n_{ijk} is the number of samples which have the jth category for node X_i and the kth configuration for its parents.

The corresponding posterior probability for GBNs is called *Bayesian Gaussian equivalent uniform* (BGeu) from Kuipers et al. (2014), which is commonly referred to as BGe. Similarly to BDe, it assumes a non-informative prior over the parameter space of each node; its expression is complicated and will not be reported here. As for CGBNs, the posterior probability combines BDe and BGe as described in Bøttcher and Dethlefsen (2003).

As a result of the difficulties outlined above, two alternatives to the use of $\Pr(\mathcal{D} \mid G)$ in structure learning have been developed. The first is the *Bayesian Information criterion* (BIC), which approximates $\Pr(\mathcal{D} \mid G)$:

$$\mathrm{BIC}(G, \mathcal{D}) \to \log \mathrm{BDe}(G, \mathcal{D}) \qquad \text{as the sample size } n \to \infty. \quad (6.16)$$

BIC is decomposable and only depends on the likelihood function,

$$\mathrm{BIC}(G, \mathcal{D}) = \sum_{i=1}^{p} \left[\log \Pr(X_i \mid \Pi_{X_i}) - \frac{|\Theta_{X_i}|}{2} \log n \right], \quad (6.17)$$

which makes it very easy to compute (see Equation (1.14) for discrete BNs, Equation (2.13) for GBNs and Equation (3.4) for CGBNs). The second alternative is to avoid the need to define a measure of goodness-of-fit for the DAG and to use conditional independence tests to learn the DAG structure. Both these approaches to structure learning will be covered in Section 6.5.1.

Once we have learned the DAG structure we can move to parameter learning, that is, we can estimate the parameters of $\mathbf{X}$. Assuming that parameters belonging to different local distributions are independent, we can estimate the parameters of one local distribution at a time. Following the Bayesian approach outlined in Equation (6.12), this would require to find the value of the Θ that maximises $\Pr(\Theta \mid G, \mathcal{D})$ through its components $\Pr(\Theta_{X_i} \mid X_i, \Pi_{X_i})$. Other approaches such as maximum likelihood and regularised estimation are also common: their advantages and disadvantages will be covered in Section 6.5.2.

Local distributions in practice involve only a small number of nodes, *i.e.*, X_i and it parents Π_{X_i}. Furthermore, their dimension usually does not scale with the number of nodes in the BN (and is often assumed to be bounded by a constant when we assume that we are working with sparse DAGs), thus avoiding the so-called *curse of dimensionality*. This means that each local distribution has a comparatively small number of parameters to estimate from the data, and that estimates are more accurate due to the better ratio between the size of each Θ_{X_i} and the sample size.

6.5.1 Structure Learning

Several algorithms have been presented in the literature for this problem, thanks to the application of results arising from probability, information and optimisation theory. Despite the (sometimes confusing) variety of theoretical backgrounds and terminology, they can all be traced to three approaches: *constraint-based*, *score-based* and *hybrid*.

All these algorithms operate under a common set of assumptions:

- There must be a one-to-one correspondence between the nodes in the DAG and the random variables in $\mathbf{X}$: this means in particular that there must not be multiple nodes which are deterministic functions of a single variable.

- All the relationships between the variables in $\mathbf{X}$ must be conditional independencies, because they are by definition the only kind of relationships that can be expressed by a BN.

- Every combination of the possible values of the variables in $\mathbf{X}$ must represent a valid, observable (even if really unlikely) event. This assumption implies a strictly positive global distribution, which is needed to have a uniquely identifiable model. Constraint-based algorithms can work even when this is not true, because the existence of a perfect map is also a sufficient condition for the uniqueness of the Markov blankets (Pearl, 1988).

- Observations are treated as independent realisations of the set of nodes. If the data present some form of temporal or spatial dependence, it must be

---

Algorithm 6.1 Inductive Causation Algorithm

1. For each pair of nodes A and B in $\mathbf{V}$ search for set $\mathbf{S}_{AB} \subset \mathbf{V}$ such that A and B are independent given $\mathbf{S}_{AB}$ and $A, B \notin \mathbf{S}_{AB}$. If there is no such a set, place an undirected arc between A and B.

2. For each pair of non-adjacent nodes A and B with a common neighbour C, check whether $C \in \mathbf{S}_{AB}$. If this is not true, set the direction of the arcs $A - C$ and $C - B$ to $A \to C$ and $C \leftarrow B$.

3. Set the direction of arcs which are still undirected by applying iteratively the following two rules:

   (a) if A is adjacent to B and there is a strictly directed path from A to B then set the direction of $A - B$ to $A \to B$;

   (b) if A and B are not adjacent but $A \to C$ and $C - B$, then change the latter to $C \to B$.

4. Return the resulting CPDAG.

---

specifically accounted for in the definition of the network, as in the dynamic BNs in Chapter 4.

6.5.1.1 *Constraint-Based Algorithms*

Constraint-based algorithms are based on the seminal work of Pearl on maps and its application to causal graphical models. His *Inductive Causation* (IC) algorithm (Verma and Pearl, 1991) provides a framework for learning the DAG structure of BNs using conditional independence tests.

The details of the IC algorithm are described in Algorithm 6.1. The first step identifies which pairs of variables are connected by an arc, regardless of its direction. These variables cannot be independent given any other subset of variables because they cannot be d-separated. This step can also be seen as a backward selection procedure starting from the saturated model with a complete graph and pruning it using statistical tests for conditional independence. The second step identifies the v-structures among all the pairs of non-adjacent nodes A and B with a common neighbour C. By definition, v-structures are the only fundamental connection in which the two non-adjacent nodes are not independent conditional on the third node. Therefore, if there is a subset of nodes that contains C and d-separates A and B, the three nodes are part of a v-structure centred on C. This condition can be verified by performing a conditional independence test for A and B against every possible subset of their common neighbours that includes C. At the end of the second step, both the skeleton and the v-structures of the network are known, so the equivalence class the BN belongs to is uniquely identified. The third and last step of the IC algorithm identifies compelled arcs and orients them iteratively to obtain the CPDAG describing the equivalence class identified by the previous steps.

A major problem of the IC algorithm is that the first two steps cannot be applied in the form described in Algorithm 6.1 to most real-world problems due to the exponential number of possible conditional independence relationships to test. This has led to the development of more efficient algorithms such as:

- *PC*: the first practical application of the IC algorithm (Spirtes et al., 2000).

- *Grow-Shrink* (GS): based on the *Grow-Shrink Markov blanket* algorithm (Margaritis, 2003), a simple forward selection Markov blanket detection approach.

- *Incremental Association* (IAMB) and its variants: based on the *Incremental Association Markov blanket* algorithm (Tsamardinos et al., 2003), a two-phase selection scheme, and its variants Fast-IAMB (Yaramakala and Margaritis, 2005) and Inter-IAMB (Tsamardinos et al., 2003).

- *MMPC* and *HITON-PC*: based on the local discovery algorithms from Tsamardinos et al. (2006) and Aliferis et al. (2010) for learning the skeleton of a BN.

- *HPC*: from the local discovery algorithm by Gasse et al. (2014).

Many of these algorithms first learn the Markov blanket of each node. This preliminary step greatly simplifies the identification of neighbours, but it does not necessarily reduce the number of conditional independence tests (and therefore of the overall computational complexity of the learning algorithm) or increase the quality of the learned networks. A thorough benchmark and review of these methods as well as those in Sections 6.5.1.2 and 6.5.1.3 can be found in Scutari et al. (2019).

Conditional independence tests used to learn discrete BNs are functions of the observed frequencies $\{n_{ijk}, i = 1, \ldots, R; j = 1, \ldots, C; k = 1, \ldots, L\}$ for the random variables X and Y and all the configurations of the conditioning variables $\mathbf{Z}$. We introduced two such tests in Section 1.5.1:

- the *mutual information* test, an information-theoretic distance measure defined as

$$\mathrm{MI}(X, Y \mid \mathbf{Z}) = \sum_{i=1}^{R} \sum_{j=1}^{C} \sum_{k=1}^{L} \frac{n_{ijk}}{n} \log \frac{n_{ijk} n_{++k}}{n_{i+k} n_{+jk}} \qquad (6.18)$$

and equivalent to the log-likelihood ratio test G^2 (they differ by a $2n$ factor, where n is the sample size);

- the classic *Pearson's* X^2 test for contingency tables,

$$\mathrm{X}^2(X, Y \mid \mathbf{Z}) = \sum_{i=1}^{R} \sum_{j=1}^{C} \sum_{k=1}^{L} \frac{(n_{ijk} - m_{ijk})^2}{m_{ijk}}, \quad \text{where} \quad m_{ijk} = \frac{n_{i+k} n_{+jk}}{n_{++k}}. \qquad (6.19)$$

Test	Asymptotic χ^2	Monte Carlo	Sequential Monte Carlo	Semiparametric χ^2
mutual information (G^2)	`"mi"`	`"mc-mi"`	`"smc-mi"`	`"sp-mi"`
Pearson's X^2	`"x2"`	`"mc-x2"`	`"smc-x2"`	`"sp-x2"`
mutual information (shrinkage)	`"mi-sh"`	–	–	–

Table 6.1
Labels of the conditional independence tests for discrete BNs implemented in
bnlearn.

Another possibility is the shrinkage estimator for the mutual information
defined by Hausser and Strimmer (2009) and studied in the context of BNs in
Scutari and Brogini (2012).

For all the tests above, the null hypothesis of independence can be tested
using:

- Their asymptotic $\chi^2_{(R-1)(C-1)L}$ distribution, which is very fast to compute
  but requires an adequate sample size to be accurate.

- The Monte Carlo permutation approach or the faster, sequential Monte
  Carlo permutation approaches described in Edwards (2000). Both
  approaches ensure that the tests are unbiased, and that they are valid even
  for small sample sizes. However, they are computationally expensive.

- The semiparametric χ^2 distribution described in Tsamardinos and
  Borboudakis (2010), which represents a compromise between the previous
  two approaches.

Several of these combinations of tests statistics and distributions are
implemented in **bnlearn**, for use both in structure learning and as standalone
tests in `ci.test` (for some examples of the latter, see Section 1.5.1). The labels
used to identify them throughout the package are reported in Table 6.1.

In the case of GBNs, conditional independence tests are functions of the
partial correlation coefficients $\rho_{XY|\mathbf{Z}}$ of X and Y given $\mathbf{Z}$. Two common
conditional independence tests are:

- the exact t test for Pearson's correlation coefficient, defined as

$$t(X, Y \mid \mathbf{Z}) = \rho_{XY|\mathbf{Z}} \sqrt{\frac{n - |\mathbf{Z}| - 2}{1 - \rho^2_{XY|\mathbf{Z}}}} \qquad (6.20)$$

and distributed as a Student's t with $n - |\mathbf{Z}| - 2$ degrees of freedom;

Test	Exact distribution	Asymptotic distribution	Monte Carlo	Sequential Monte Carlo
t test	`"cor"`	–	`"mc-cor"`	`"smc-xcor"`
Fisher's Z test	–	`"zf"`	`"mc-zf"`	`"smc-zf"`
mutual information	–	`"mi-g"`	`"mc-mi-g"`	`"smc-mi-g"`
mutual information (shrinkage)	–	`"mi-g-sh"`	–	–

Table 6.2
Labels of the conditional independence tests for GBNs implemented in **bnlearn**.

- *Fisher's Z test*, a transformation of $\rho_{XY|\mathbf{Z}}$ with an asymptotic normal distribution and defined as

$$Z(X, Y \mid \mathbf{Z}) = \log\left(\frac{1 + \rho_{XY|\mathbf{Z}}}{1 - \rho_{XY|\mathbf{Z}}}\right)\frac{\sqrt{n - |\mathbf{Z}| - 3}}{2}, \qquad (6.21)$$

where n is the number of observations and $|\mathbf{Z}|$ is the number of nodes in $\mathbf{Z}$.

Another possible choice is the mutual information test defined in Kullback (1968), which is again proportional to the corresponding log-likelihood ratio test and has an asymptotic χ_1^2 distribution. The shrinkage estimator for the covariance matrix developed by Shäfer and Strimmer (2005) can also be used to construct a regularised mutual information test, again with a χ_1^2 asymptotic distribution. The labels used to identify these tests and their Monte Carlo variants in **bnlearn** are reported in Table 6.2.

As for CGBNs, the only conditional independence test available in **bnlearn** is the mutual information test, labelled `"mi-cg"`.

Constraint-based structure learning algorithms are implemented in **bnlearn** in the gs, iamb, fast.iamb, inter.iamb, mmpc, si.hiton.pc and hpc functions. All these functions accept the following arguments:

- x, the data the network will be learned from.

- whitelist and blacklist, to force the inclusion or the exclusion of specific arcs from the network (examples in Sections 4.4 and 2.5.1).

- test, the label of the conditional independence test to be used. It defaults to `"mi"` for discrete BNs, `"cor"` for GBNs and `"mi-cg"` for CGBNs.

- alpha, the type I error threshold for the test. It defaults to $\alpha = 0.05$, as is common in the literature. Note that, regardless of the dimension of the BN, there is no need to apply any multiple testing adjustment to α, because constraint-based learning algorithms are largely self-adjusting in that respect.

- B, the number of permutations for Monte Carlo tests.

- debug, which prints the sequence of tests performed by the learning algorithm. Setting it to TRUE may be useful for understanding how structure learning works in practice, for debugging, and also for assessing the swiftness of the learning algorithm.

Consider one more time the example of structure learning from the cropdata200 data set presented in Section 2.5, which is based on 200 observations. By default, Grow-Shrink uses the exact t test for partial correlations with $\alpha = 0.05$.

```
bn.cor <- gs(cropdata200, test = "cor", alpha = 0.05)
modelstring(bn.cor)
 [1] "[E][G][N][V|E:G][W|V][C|N:W]"
```

The small sample size seems to reduce the power of the tests, and the arc $V \rightarrow N$ is missing from the DAG as a result. Therefore, we may wonder whether changing the conditional independence test to Fisher's Z test or to a Monte Carlo test may help in learning the correct structure.

```
bn.zf <- gs(cropdata200, test = "zf", alpha = 0.05)
bn.mc <- gs(cropdata200, test = "mc-cor", B = 1000)
all.equal(bn.cor, bn.zf)
 [1] TRUE
all.equal(bn.cor, bn.mc)
 [1] TRUE
```

As we can see for the output above, both tests result in the same DAG as the exact t test. Likewise, using IAMB instead of Grow-Shrink does not result in the correct structure.

```
bn.iamb <- iamb(cropdata200, test = "cor", alpha = 0.05)
all.equal(bn.cor, bn.iamb)
 [1] TRUE
```

Therefore, we may presume that the problems in learning $V \rightarrow N$ lie more in the small sample size than in the limitations of a particular algorithm or a particular test. If we set the debug argument to TRUE, we can identify exactly which tests fail to detect the dependence.

```
gs(cropdata200, test = "cor", alpha = 0.05, debug = TRUE)
 [...]
 * learning the markov blanket of N .
 [...]
   * checking node V for inclusion.
```

```
      > N indep. V given ' C ' (p-value: 0.3048644).
   * checking node W for inclusion.
      > node W included in the markov blanket (p-value: 7.89e-07).
      > markov blanket ( 2 nodes ) now is ' C W '.
      > restarting grow loop.
   * checking node V for inclusion.
      > N indep. V given ' C W ' (p-value: 0.1416876).
   * checking node W for exclusion (shrinking phase).
      > node W remains in the markov blanket. (p-value: 7.89e-07)
[...]
```

When learning the Markov blanket of N, we have that node C is included before V. Apparently, the dependence between C and N is much stronger than the dependence between V and N; the latter is not statistically significant when the former is taken into account. As a result, the marginal dependence test between V and N is not significant at $\alpha = 0.05$ while the same test conditional on C is significant.

```
ci.test("N", "V", test = "cor", data = cropdata200)

          Pearson's Correlation

 data:  N ~ V
 cor = -0.06, df = 198, p-value = 0.4
 alternative hypothesis: true value is not equal to 0
ci.test("N", "V", "C", test = "cor", data = cropdata200)

          Pearson's Correlation

 data:  N ~ V | C
 cor = -0.2, df = 197, p-value = 0.002
 alternative hypothesis: true value is not equal to 0
```

Now that we have diagnosed the problem behind the missing V → N arc, we can include it using the whitelist argument and thus obtain the correct DAG (pdag2 in Section 2.5).

```
bn.cor <- gs(cropdata200, test = "cor", alpha = 0.05,
           whitelist = c("V", "N"))
all.equal(bn.cor, dag.bnlearn)
 [1] TRUE
```

6.5.1.2 Score-Based Algorithms

Score-based learning algorithms represent the application of different optimisation techniques to the problem of learning the structure of a BN. Each candidate BN is assigned a *network score* reflecting its goodness of fit, which

---

Algorithm 6.2 Hill-Climbing Algorithm

---

1. Choose a network structure G over $\mathbf{V}$, usually (but not necessarily) empty.

2. Compute the score of G, denoted as $\text{Score}_G = \text{Score}(G)$.

3. Set $maxscore = \text{Score}_G$.

4. Repeat the following steps as long as $maxscore$ increases:

    (a) for every possible arc addition, deletion or reversal not resulting in a cyclic network:

        i. compute the score of the modified network G^*,
           $\text{Score}_{G^*} = \text{Score}(G^*)$:

        ii. if $\text{Score}_{G^*} > \text{Score}_G$, set $G = G^*$ and $\text{Score}_G = \text{Score}_{G^*}$.

    (b) update $maxscore$ with the new value of Score_G.

5. Return the DAG G.

---

the algorithm then attempts to maximise. This can be done in a *heuristic* way, to achieve both speed and ease of implementation at the cost of losing any guarantee that the algorithm will identify a global optimum; or in an *exact* way, which is guaranteed to find the best DAG for the given $\mathcal{D}$ but at a significant computational cost. Some common heuristic algorithms are:

- *Greedy search* algorithms such as *hill-climbing* with *random restarts* or *tabu search* (Bouckaert, 1995). These algorithms explore the search space starting from a network structure (usually without any arc) and adding, deleting or reversing one arc at a time until the score can no longer be improved (see Algorithm 6.2).

- *Genetic* algorithms, which mimic natural evolution through the iterative selection of the "fittest" models and the hybridisation of their characteristics (Larrañaga et al., 1997). In this case the search space is explored through the *crossover* (which combines the structure of two networks) and *mutation* (which introduces random alterations) stochastic operators.

- *Simulated annealing* (Bouckaert, 1995). This algorithm performs a stochastic local search by accepting changes that increase the network score and, at the same time, allowing changes that decrease it with a probability inversely proportional to the score decrease.

A comprehensive review of these heuristics, as well as related approaches from the field of artificial intelligence, is provided in Russell and Norvig (2009). Exact algorithms, on the other hand, focus more on how to not the whole explore space of the DAGs by ruling out candidate DAGs without having to score them. Notable examples include Cussens (2012), Suzuki (2017) and

Koivisto and Sood (2004): they use a variety of constraints and bounds to rule out candidate DAGs extrapolating for the scores of the DAGs that have already been scored.

As far as network scores are concerned, the only two options in common use are those we introduced in Section 6.5: posterior probabilities arising from flat priors such as BDe (for discrete BNs) and BGe (for GBNs), and the BIC score. In **bnlearn**, they are labelled "bde", "bge", "bic" (for discrete BNs), "bic-g" (for GBNs) and "bic-cg" (for CGBNs). All these score assign the values to DAGs in the same equivalence class, which means that in practice they can only learn the CPDAG for that class; this property is called *score equivalence*. This lack of options, compared to the number of conditional independence tests we covered in the previous section, can be attributed to the difficulty of specifying a tractable likelihood or posterior distribution for the DAG.

As an example of score-based structure learning, consider once again the structure learning example from Section 1.5. After considering some manual arc additions and removal, we called the hill-climbing implementation in **bnlearn** and obtained a DAG with a better BIC score.

```
learned <- hc(survey, score = "bic")
modelstring(learned)
[1] "[R][E|R][T|R][A|E][O|E][S|E]"
score(learned, data = survey, type = "bic")
[1] -1998.43
```

The tabu function, which implements tabu search, returns the same DAG. Both hc and tabu have argument sets similar to those of the functions presented in the previous section: the main difference is the presence of a score argument instead of test, alpha and B. Hence we can investigate the workings of hc by setting the debug argument to TRUE, like we did for gs in the previous section.

```
learned <- hc(survey, score = "bic", debug = TRUE)
* starting from the following network:

  Random/Generated Bayesian network

  model:
   [A][R][E][O][S][T]
  nodes:                             6
  arcs:                              0
    undirected arcs:                 0
    directed arcs:                   0
  average markov blanket size:       0.00
  average neighbourhood size:        0.00
  average branching factor:          0.00

  generation algorithm:              Empty
```

```
* current score: -2008.943
[...]
* best operation was: adding R -> E .
* current network is :
[...]
* best operation was: adding E -> S .
[...]
* best operation was: adding R -> T .
[...]
* best operation was: adding E -> A .    `
[...]
* best operation was: adding E -> O .
[...]
```

The search for the network with the optimal score starts, by default, from the empty DAG. The operation that increases the BIC score the most at each step is the addition of one of the arcs that will be present in final DAG (see Figure 6.5).

Neither hc nor tabu are able to learn the true DAG. There are many reasons for such a behaviour. For instance, it is possible for both algorithms to get stuck at a local maximum because of an unfavourable choice for the starting point of the search. This does not appear to be the case for the survey data, because even when starting from the true DAG (survey.dag) the final DAG is not the true one.

```
survey.dag <- model2network("[A][S][E|A:S][O|E][R|E][T|O:R]")
learned.start <- hc(survey, score = "bic", start = survey.dag)
modelstring(learned.start)
 [1] "[S][E|S][A|E][O|E][R|E][T|R]"
all.equal(cpdag(learned), cpdag(learned.start))
 [1] TRUE
```

As an alternative, we could also start the search from a randomly-generated graph.

```
hc(survey, score = "bic", start = random.graph(names(survey)))
```

Furthermore, it is interesting to note that the networks returned for the different starting DAGs fall in the same equivalence class. This suggests that the dependencies specified by survey.dag are not completely supported by the survey data, and that hc and tabu do not seem to be affected by convergence or numerical problems.

6.5.1.3 Hybrid Algorithms

Hybrid learning algorithms combine constraint-based and score-based algorithms to offset the respective weaknesses and produce reliable network structures in a wide variety of situations. The two best-known members of this

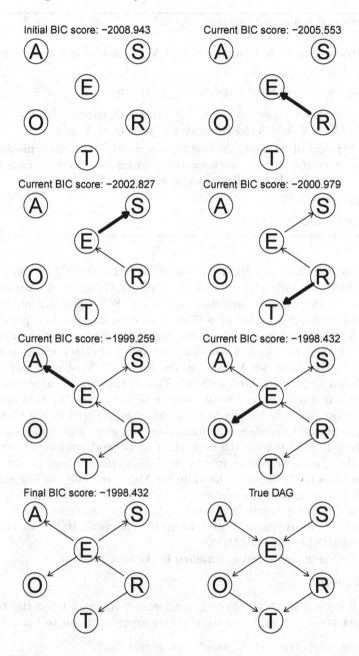

Figure 6.5
Steps performed by hill-climbing when learning the DAG with optimal BIC score from the survey data, from top to bottom. The DAG in the bottom right panel is the original DAG from Chapter 1.

Algorithm 6.3 Sparse Candidate Algorithm

---

1. Choose a network structure G over $\mathbf{V}$, usually (but not necessarily) empty.

2. Repeat the following steps until convergence:

   (a) **restrict:** select a set $\mathbf{C}_i$ of candidate parents for each node $X_i \in \mathbf{V}$, which must include the parents of X_i in G;

   (b) **maximise:** find the network structure G^* that maximises $\text{Score}(G^*)$ among the networks in which the parents of each node X_i are included in the corresponding set $\mathbf{C}_i$;

   (c) set $G = G^*$.

3. Return the DAG G.

---

family are the *Max-Min Hill-Climbing* algorithm (MMHC) by Tsamardinos et al. (2006) and the *Hybrid HPC* (H^2PC) by Gasse et al. (2014), which improve on the *Sparse Candidate* algorithm (SC) originally proposed by Friedman et al. (1999). MMHC is illustrated as an example in Algorithm 6.3.

Both these algorithms are based on two steps, called *restrict* and *maximise*. In the first step, the candidate set for the parents of each node X_i is reduced from the whole node set $\mathbf{V}$ to a smaller set $\mathbf{C}_i \subset \mathbf{V}$ of nodes which have been shown to be associated with X_i. This in turn results in a smaller and more regular search space. The second step seeks the network that maximises a given score function, subject to the constraints imposed by the $\mathbf{C}_i$ sets.

In the Sparse Candidate algorithm these two steps are applied iteratively until there is no change in the network or no network improves the network score; the choice of the heuristics used to perform them is left to the implementation. On the other hand, in the Max-Min Hill-Climbing algorithm *restrict* and *maximise* are performed only once: the *Max-Min Parents and Children* (MMPC) heuristic is used to learn the candidate sets $\mathbf{C}_i$ and a hill-climbing greedy search to find the optimal network. H^2PC does the same but using HPC instead of MMPC.

MMHC is implemented in **bnlearn** in the mmhc function,

```
mmhc(survey)
```

which is equivalent to the more general rsmax2 function when the restrict argument is set to "mmpc" and the maximize argument is set to "hc".

```
rsmax2(survey, restrict = "mmpc", maximize = "hc")
```

rsmax2 implements a single-iteration variant of the Sparse Candidate algorithm; the algorithms used in each step can be specified independently. Suppose, for example, that we would like to use Hiton-PC with Pearson's X^2 test in the restrict step and tabu search with the BDe score and an imaginary sample size of 1 in the maximise step. They are, after all, among the best algorithms in the respective groups. We can do that as follows.

```
rsmax2(survey, restrict = "si.hiton.pc", maximize = "tabu",
  restrict.args = list(test = "x2"),
  maximize.args = list(score = "bde", iss = 1))
```

Clearly, even though any combination of constraint-based algorithms (for the *restrict* step), score-based algorithms (for the *maximise* step), conditional independence tests and network scores will work, some make more sense than others.

6.5.2 Parameter Learning

Once the structure of the BN has been learned from the data, the task of estimating and updating the parameters of the global distribution is greatly simplified by the decomposition into local distributions. Two approaches are common in the literature: *maximum likelihood* and *Bayesian estimation*. Examples of both were covered in Section 1.4 (for discrete BNs) Section 2.4 (for GBNs) and Section 3.3 (for CGBNs). Other choices, such as the shrinkage estimators presented in Hausser and Strimmer (2009) and Shäfer and Strimmer (2005) are certainly possible. It is important to note that the approach used to learn the structure of the BNs does necessarily determine which approaches can be used in parameter learning. For instance, using posterior densities in both structure and parameter learning makes the interpretation of the BN and its inference straightforward. However, using a mutual information test for structure learning and posterior estimates for parameter learning is also common, as is using shrinkage approaches and maximum likelihood parameter estimates.

Even though local distributions in practice involve only a small number of variables, and their dimension usually does not scale with the size of the BN, parameter estimation is still problematic in some situations. For example, it is common to have sample sizes much smaller than the number of variables included in the model. This is typical of high-throughput biological data sets, such as those arising from whole-genome and omics assays. In this setting, which is called "small n, large p", estimates have a high variability unless particular care is taken in both structure and parameter learning.

6.6 Bayesian Network Inference

Learning the structure and the parameters of a BN can provide significant insights into the nature of the data. For instance, it can highlight the dependence structure of the data and, under the right conditions, the causal structure as well. Furthermore, the parameters associated with each node provide a concise description of that node's behaviour relative to its parents. However, there are many questions that can only be answered by incorporating new evidence in the BN or by investigating the probability of complex events. Several examples of such questions can be found in Sections 1.6 (for discrete BNs), 2.6 (for GBNs), 3.5 (for CGBNs) and 4.5 (for dynamic BNs).

6.6.1   Probabilistic Reasoning and Evidence

BNs, like other statistical models, can be used to answer questions about the nature of the data that go beyond the mere description of the behaviour of the observed sample. The techniques used to obtain those answers are known in general as *inference*. For BNs, the process of answering these questions is also known as *probabilistic reasoning* or *belief updating*, while the questions themselves are called *queries*. Both names were introduced by Pearl (1988) and borrowed from expert systems theory (*e.g.*, you would submit a *query* to an *expert* to get an opinion, and *update your beliefs* given new evidence), and have completely replaced traditional statistical terminology in reference books such as Koller and Friedman (2009).

In practice, probabilistic reasoning on BNs works in the framework of Bayesian statistics and focuses on the computation of posterior probabilities or densities. For example, suppose we have learned a BN $\mathcal{B}$ with structure G and parameters Θ under one of the distributional assumptions we explored in previous chapters. Subsequently, we want to use $\mathcal{B}$ to investigate the effects of a new piece of *evidence* $\mathbf{E}$ using the knowledge encoded in $\mathcal{B}$, that is, to investigate the posterior distribution $\Pr(\mathbf{X} \mid \mathbf{E}, \mathcal{B}) = \Pr(\mathbf{X} \mid \mathbf{E}, G, \Theta)$.

The approaches used for this kind of analysis vary depending on the nature of $\mathbf{E}$ and on the nature of information we are interested in. The two most common kinds of evidence are:

- *Hard evidence*, an instantiation of one or more variables in the network. In other words,

$$\mathbf{E} = \{X_{i_1} = e_1, X_{i_2} = e_2, \ldots, X_{i_k} = e_k\}, \quad i_1 \neq \ldots \neq i_k \in \{1, \ldots n\}, \tag{6.22}$$

  which ranges from the value of a single variable X_i to a complete specification for $\mathbf{X}$. Such an instantiation may come, for instance, from a new (partial or complete) observation recorded after $\mathcal{B}$ was learned.

- *Soft evidence*, a new distribution for one or more variables in the network. Since both the network structure and the distributional assumptions are treated as fixed, soft evidence is usually specified as a new set of parameters,

$$\mathbf{E} = \left\{ X_{i_1} \sim (\Theta_{X_{i_1}}), X_{i_2} \sim (\Theta_{X_{i_2}}), \ldots, X_{i_k} \sim (\Theta_{X_{i_k}}) \right\}. \tag{6.23}$$

  This new distribution may be, for instance, the null distribution in a hypothesis testing problem: the use of a specific conditional probability table in case of discrete BNs or a zero value for some regression coefficients in case of GBNs or CGBNs.

As far as queries are concerned, we will focus on *conditional probability* (CPQ) and *maximum a posteriori* (MAP) queries, also known as *most probable explanation* (MPE) queries. Both apply mainly to hard evidence, even though they can be used in combination with soft evidence.

Conditional probability queries are concerned with the distribution of a subset of variables $\mathbf{Q} = \{X_{j_1}, \ldots, X_{j_l}\}$ given some hard evidence $\mathbf{E}$ on another set $X_{i_1}, \ldots, X_{i_k}$ of variables in $\mathbf{X}$. The two sets of variables can be assumed to be disjoint. The distribution we are interested in is

$$\mathrm{CPQ}(\mathbf{Q} \mid \mathbf{E}, \mathcal{B}) = \Pr(\mathbf{Q} \mid \mathbf{E}, G, \Theta) = \Pr(X_{j_1}, \ldots, X_{j_l} \mid \mathbf{E}, G, \Theta), \qquad (6.24)$$

which is the marginal posterior probability distribution of $\mathbf{Q}$, that is,

$$\Pr(\mathbf{Q} \mid \mathbf{E}, G, \Theta) = \int \Pr(\mathbf{X} \mid \mathbf{E}, G, \Theta) \, d(\mathbf{X} \setminus \mathbf{Q}). \qquad (6.25)$$

This class of queries has many useful applications due to their versatility. For instance, it can be used to assess the interaction between two sets of experimental design factors for a trait of interest: the latter would be considered as the hard evidence E, while the former would play the role of the set of query variables $\mathbf{Q}$. As another example, the odds of an unfavourable outcome $\mathbf{Q}$ can be assessed for different sets of hard evidence $\mathbf{E}_1, \mathbf{E}_2, \ldots, \mathbf{E}_m$.

Maximum a posteriori queries are concerned with finding the configuration $\mathbf{q}^*$ of the variables in $\mathbf{Q}$ that has the highest posterior probability (for discrete BNs) or the maximum posterior density (for GBNs, CGBNs and general BNs),

$$\mathrm{MAP}(\mathbf{Q} \mid \mathbf{E}, \mathcal{B}) = \mathbf{q}^* = \underset{\mathbf{q}}{\arg\max} \Pr(\mathbf{Q} = \mathbf{q} \mid \mathbf{E}, G, \Theta). \qquad (6.26)$$

Applications of this kind of query fall into two categories: imputing missing data from partially observed hard evidence, where the variables in $\mathbf{Q}$ are not observed and are to be imputed from those in $\mathbf{E}$, or comparing $\mathbf{q}^*$ with the observed values for the variables in $\mathbf{Q}$ for completely observed hard evidence. Prediction in discrete BNs can also be thought of as a MAP query; on the other hand, prediction in GBNs (or continuous nodes in CGBNs is computed using posterior expectations rather than MAPs.

Both conditional probability queries and maximum a posteriori queries can also be used with soft evidence, albeit with different interpretations. When $\mathbf{E}$ encodes hard evidence it is not stochastic; it is an observed value. In this case, $\Pr(\mathbf{Q} = \mathbf{q} \mid \mathbf{E}, G, \Theta)$ is not stochastic. However, when $\mathbf{E}$ encodes soft evidence it is still a random variable, and as a result $\Pr(\mathbf{Q} = \mathbf{q} \mid \mathbf{E}, G, \Theta)$ is stochastic as well. Therefore, the answers provided by the queries described in this section must be manipulated and evaluated according to the nature of the evidence they are based on.

It must be underlined that all techniques presented in this section are applications of the same basic principle: we modify the joint probability distribution of the nodes to incorporate a new piece of information. In the case of hard evidence, the distribution is conditioned on the values of some nodes; in the case of soft evidence, some local distributions are modified.

6.6.2 Algorithms for Belief Updating

The estimation of the posterior probabilities and densities in the previous section is a fundamental problem in the evaluation of queries. Queries involving very small probabilities or large networks are particularly problematic even with the best algorithms available in the literature, because they present both computational and probabilistic challenges.

Algorithms for belief updating can be classified as either *exact* or *approximate*. Both build upon the fundamental properties of BNs introduced in Section 6.2 to avoid the curse of dimensionality through the use of *local computations*, that is, by only using local distributions. So, for instance, the marginalisation in Equation (6.25) can be rewritten as

$$\Pr(\mathbf{Q} \mid \mathbf{E}, G, \Theta) = \int \Pr(\mathbf{X} \mid \mathbf{E}, G, \Theta) \, d(\mathbf{X} \setminus \mathbf{Q})$$

$$= \int \left[\prod_{i=1}^{p} \Pr(X_i \mid \mathbf{E}, \Pi_{X_i}, \Theta_{X_i}) \right] d(\mathbf{X} \setminus \mathbf{Q})$$

$$= \prod_{i:X_i \in \mathbf{Q}} \int \Pr(X_i \mid \mathbf{E}, \Pi_{X_i}, \Theta_{X_i}) \, dX_i. \tag{6.27}$$

The link between d-separation and conditional independence can also be used to further reduce the dimension of the problem. From Definition 6.2, variables that are d-separated from $\mathbf{Q}$ by $\mathbf{E}$ cannot influence the outcome of the query. Therefore, they may be completely disregarded in computing posterior probabilities.

6.6.2.1 *Exact Inference Algorithms*

Exact inference algorithms combine repeated applications of Bayes' theorem with local computations to compute the exact value $\Pr(\mathbf{Q} \mid \mathbf{E}, G, \Theta)$. Due to their nature, such algorithms are feasible only for small or very simple networks, such as trees and polytrees, or for sparse networks that have few arcs compared to the number of nodes. In the worst case, their computational complexity is exponential in the number of variables.

The two best-known exact inference algorithms are *variable elimination* and belief updates based on *junction trees*. Both were originally derived for discrete BNs, and have later been extended to continuous and hybrid networks. Variable elimination uses the structure of the BN directly, specifying the optimal sequence of operations on the local distributions and how to cache

---

Algorithm 6.4 Junction Tree Clustering Algorithm

---

1. **Moralise:** create the moral graph of the BN $\mathcal{B}$ as explained in Section 6.4.

2. **Triangulate:** break every cycle spanning 4 or more nodes into subcycles of exactly three nodes by adding arcs to the moral graph, thus obtaining a *triangulated graph*.

3. **Cliques:** identify the *cliques* $C_1, \ldots, C_k$ of the triangulated graph, *i.e.*, maximal subsets of nodes in which each element is adjacent to all the others.

4. **Junction Tree:** create a tree in which each clique is a node, and adjacent cliques are linked by arcs. The tree must satisfy the *running intersection property*: if a node belongs to two cliques C_i and C_j, it must be also included in all the cliques in the (unique) path that connects C_i and C_j.

5. **Parameters:** use the parameters of the local distributions of $\mathcal{B}$ to compute the parameter sets of the compound nodes of the junction tree.

---

intermediate results to avoid unnecessary computations. An alternative is to transform the BN into a junction tree as a preprocessing step, and use that to perform belief propagation. As the name suggests, a junction tree is a transformation that reduces any network structure into a tree whose nodes are clusters of the original nodes. Belief updates can be performed efficiently on this simplified representation of the BN using Kim and Pearl's Message Passing algorithm. This second approach can be used for graphical models other than BNs; it is more general in its formulation and thus more commonly found implemented in software. For this reason we cover junction trees below, and we refer the reader to Korb and Nicholson (2011) and Koller and Friedman (2009) for an exhaustive explanation and step-by-step examples of variable elimination.

The steps required to construct junction trees are described in Algorithm 6.4. Consider again, as an example, one of the queries we explored in Section 1.6.2.2: assessing from our survey the probability to find a man driving a car given that he has a high school education.

In step 1 we construct the moral graph (Figure 6.6, top left) of the BN as in Section 6.4.

```
survey.moral = moral(survey.dag)
```

Then in step 2 we *triangulate* the moral graph so that the longest (now undirected) cycle is at most 3-nodes long. For reasons that are beyond the

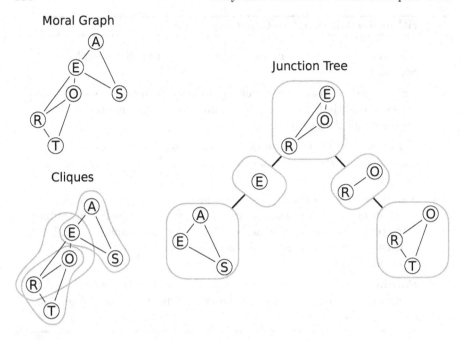

Figure 6.6
The moral graph, the cliques and the junction tree constructed from
survey.dag to perform exact inference using Algorithm 6.4.

scope of this book,[1] this property allows the global probability distribution
of **X** to decompose over the nodes of the junction tree. In other words, if
we find a cycle like X1 — X2 — X3 — X4 — X1 spanning four nodes, but neither
X1 — X3 nor X2 — X4 are present in the moral graph, we add one of these two
arcs in order to cut the cycle into two shorter cycles spanning three nodes.
If we choose to add X1 — X3, we obtain X1 — X2 — X3 and X1 — X3 — X4; if we
choose to add X2 — X4 we obtain X1 — X2 — X4 and X2 — X3 — X4. In the case
of survey.moral, the longest cycle is already shorter than four nodes so there
nothing to do for this step.

From this graph, in step 3 we identify *cliques* of nodes that are all
connected with each other. The nodes in each clique are therefore all
dependent on each other, and it makes intuitive sense to treat them as
a single, multivariate random variable instead of a collection of distinct
univariate variables. From Figure 6.6 (bottom left), we can see three cliques
in moral.graph:

$$C_1 = \{\text{A}, \text{E}, \text{S}\}, \qquad C_2 = \{\text{E}, \text{O}, \text{R}\}, \qquad C_3 = \{\text{O}, \text{R}, \text{T}\}.$$

---

[1]For the interested reader, see Castillo et al. (1997). In short, a triangulated moral graph
is effectively a decomposable Markov network.

The intersections between the cliques,

$$S_{12} = \{E\} \qquad \text{and} \qquad S_{23} = \{O, R\},$$

are called *separators* and will also be used as nodes in the junction tree to connect the cliques. This is the tree that is shown in Figure 6.6 (right) and that is described in step 4. Arcs are represented as undirected because their direction is irrelevant for subsequent computations: we could choose any of C_1, C_2, C_3 as the root node and set arc directions accordingly. In addition, it is important to note that in general we can construct multiple valid junction trees from the same triangulated moral graph: intersections between cliques may determine a graph of connections that we can prune down to a tree in different ways.

Finally, in step 5 we compute the distribution and the parameters of each clique. These are known as *potentials* and differ from probability distributions in that they are not required to be normalised to sum up or to integrate to 1; a single normalisation can be performed at the end of belief propagation in order to compute the result of our query. If bn is the bn.fit object containing the survey BN, $\Pr(C_1) = \Pr(A, S, E) = \Pr(E \mid A, S) \Pr(A) \Pr(S)$ which are the local distributions of A, S, E.

```
C1 <- coef(survey.bn$E)
for (a in A.lv)
  for (s in S.lv)
    C1[, a, s] <- C1[, A = a, S = s] *
                    coef(survey.bn$A)[a] * coef(survey.bn$S)[s]
S12 <- margin.table(C1, 1)
```

The distribution $\Pr(S_{12}) = \Pr(E)$ is just the marginal of $\Pr(C_1)$ over A and S, which can be easily obtained with margin.table.

Similarly, $\Pr(C_2) = \Pr(E, O, R) = \Pr(O \mid E) \Pr(R \mid E) \Pr(E)$ where $\Pr(E) = \Pr(S_{12})$ from above and $\Pr(O \mid E)$, $\Pr(R \mid E)$ are the local distributions of O and E. $\Pr(S_{23})$ can be computed from $\Pr(C_2)$ with margin.table as before.

```
C2 <- array(0, dim = c(2, 2, 2),
        dimnames = list(O = O.lv, R = R.lv, E = E.lv))
for (o in O.lv)
  for (r in R.lv)
    for (e in E.lv)
      C2[o, r, e] <- coef(survey.bn$O)[o, e] *
                    coef(survey.bn$R)[r, e] * S12[e]
S23 <- margin.table(C2, 1:2)
```

Finally, $\Pr(C_3) = \Pr(O, R, T) = \Pr(T \mid O, R) \Pr(O, R)$ where $\Pr(O, R) = \Pr(S_{23})$ and $\Pr(T \mid O, R)$ is the local distribution of T.

```
C3 <- coef(survey.bn$T)
```

```
for (t in T.lv)
  for (o in O.lv)
    for (r in R.lv)
      C3[t, o, r] <- C3[t, o, r] * S23[o, r]
```

This completes the construction of the junction tree: we now have both its structure and its parameters. In order to answer our query we need to incorporate the evidence (E is equal to "high"), that is, we need to modify the probabilities in the junction tree to reflect that we assume $\Pr(\mathsf{E} = \mathsf{high}) = 1$. E appears in C_1, so we update its distribution from $\Pr(C_1)$ to $\Pr(C_1^*)$

$$\Pr(C_1) = \Pr(\mathsf{A}, \mathsf{S} \mid \mathsf{E}) \Pr(\mathsf{E}) \longrightarrow \widetilde{\Pr}(C_1) = \Pr(\mathsf{A}, \mathsf{S} \mid \mathsf{E})\widetilde{\Pr}(\mathsf{E})$$

by factorising and then replacing the original $\Pr(\mathsf{E})$ with $\widetilde{\Pr}(\mathsf{E} = \mathsf{high}) = 1$ and $\widetilde{\Pr}(\mathsf{E} = \mathsf{uni}) = 0$.

```
new.C1 <- C1
for (e in E.lv)
  for (a in A.lv)
    for (s in S.lv)
      new.C1[e, a, s] <-
        C1[e, a, s] / S12[e] * ifelse(e == "high", 1, 0)
```

The clique new.C1 is now *updated* with our *beliefs* as described in the query, and we *propagate* this update to the other cliques and separators in the junction tree. The distribution of S_{12} is then

```
new.S12 <- margin.table(new.C1, 1)
```

which is trivially equal to the evidence since $S_{12} = \{\mathsf{E}\}$. The flow of information from the updated C_1 to S_{12} is the *message* that is *passed* between them in Kim and Pearl's algorithm.

Updating C_2 likewise means factoring $\Pr(\mathsf{E})$ out of $\Pr(\mathsf{E}, \mathsf{O}, \mathsf{R})$ and replacing it with $\widetilde{\Pr}(\mathsf{E})$ from the updated S_{12}; and calling margin.table on the resulting new.C2 will give the updated new.S23.

```
new.C2 <- C2
for (o in O.lv)
  for (r in R.lv)
    for (e in E.lv)
      new.C2[o, r, e] <- C2[o, r, e] / S12[e] * new.S12[e]
new.S23 <- margin.table(new.C2, 1:2)
```

Finally, we can update C_3 by factoring $\Pr(\mathsf{O}, \mathsf{R})$ out of $\Pr(\mathsf{O}, \mathsf{R}, \mathsf{T})$ and replacing it with the updated $\widetilde{\Pr}(\mathsf{O}, \mathsf{R})$ from the updated S_{23}.

```
new.C3 <- C3
for (t in T.lv)
  for (o in O.lv)
```

Algorithm 6.5 Logic Sampling Algorithm

---

1. Order the variables in $\mathbf{X}$ according to the topological partial ordering implied by G, say $X_{(1)} \prec X_{(2)} \prec \ldots \prec X_{(p)}$.

2. Set $n_{\mathbf{E}} = 0$ and $n_{\mathbf{E},\mathbf{q}} = 0$.

3. For a suitably large number of samples $\mathbf{x} = (x_1, \ldots, x_p)$:

    (a) generate $x_{(i)}, i = 1, \ldots, p$ from $X_{(i)} \mid \Pi_{X_{(i)}}$ taking advantage of the fact that, thanks to the topological ordering, by the time we are considering X_i we have already generated the values of all its parents $\Pi_{X_{(i)}}$;

    (b) if $\mathbf{x}$ includes $\mathbf{E}$, set $n_{\mathbf{E}} = n_{\mathbf{E}} + 1$;

    (c) if $\mathbf{x}$ includes both $\mathbf{Q} = \mathbf{q}$ and $\mathbf{E}$, set $n_{\mathbf{E},\mathbf{q}} = n_{\mathbf{E},\mathbf{q}} + 1$.

4. Estimate $\Pr(\mathbf{Q} \mid \mathbf{E}, G, \Theta)$ with $n_{\mathbf{E},\mathbf{q}}/n_{\mathbf{E}}$.

```
for (r in R.lv)
    new.C3[t, o, r] <- C3[t, o, r] / S23[o, r] * new.S23[o, r]
```

Computing $\Pr(\mathsf{S} = \mathsf{M}, \mathsf{T} = \mathsf{car})$ at this point can be done easily by:

```
T <- margin.table(new.C3, 1)
S <- margin.table(new.C1, 3)
as.numeric(S["M"] * T["car"])
 [1] 0.3427
```

because Sex and Transportation are in different cliques and are separated by Education, hence they are independent. The value we obtain is identical to that computed by querygrain on page 24, as expected.

6.6.2.2 *Approximate Inference Algorithms*

Approximate inference algorithms use Monte Carlo simulations to sample from the global distribution of $\mathbf{X}$ and thus estimate $\Pr(\mathbf{Q} \mid \mathbf{E}, G, \Theta)$. In particular, they generate a large number of samples from $\mathcal{B}$ and they estimate the relevant conditional probabilities by weighting the samples that include both $\mathbf{E}$ and $\mathbf{Q} = \mathbf{q}$ against those that include only $\mathbf{E}$. In computer science, these random samples are often called *particles*, and the algorithms that make use of them are known as *particle filters* or *particle-based methods*.

The approaches used for both random sampling and weighting vary greatly, and their combination has resulted in several approximate algorithms being proposed in the literature. Random sampling ranges from the generation of independent samples to more complex MCMC schemes, such as those we explored in Chapter 5. Common choices are either *rejection sampling* or *importance sampling*. Furthermore, weights range from uniform to likelihood functions to various estimates of posterior probability. The simplest combination of these two classes of approaches is known as either *forward* or *logic sampling*: it is described in Algorithm 6.5 and illustrated in detail in both Korb and Nicholson (2011) and Koller and Friedman (2009). Logic sampling combines rejection sampling and uniform weights, essentially counting the proportion of generated samples including **E** that also include **Q** = **q**.

Consider once more the query we just answered using junction trees, this time with cpquery (using method = "ls" for logic sampling, which is the default method).

```
cpquery(survey.bn, event = (S == "M") & (T == "car"),
        evidence = (E == "high"), n = 10^6, method = "ls")
[1] 0.34345
```

The first part of Algorithm 6.5 concerns the creation of the particles. In **bnlearn**, the sampling of random observations from a BN is implemented in the rbn function, which takes a bn.fit object and the number of samples to generate as arguments.

```
particles <- rbn(survey.bn, 10^6)
head(particles, n = 5)
          A     E   O     R S     T
1 adult high emp     big M other
2 adult high emp     big F train
3 adult high emp     big M other
4 young high emp small F train
5 young high emp     big M   car
```

Calling rbn corresponds to steps 1 and 3(a). Step 3(b) requires finding the particles that match the evidence, E == "high", and counting them to obtain $n_{\mathbf{E}}$.

```
partE <- particles[(particles[, "E"] == "high"), ]
nE <- nrow(partE)
```

Similarly, step 3(c) requires finding in partE those particles that also match the event we are investigating, (S == "M") & (T == "car"), and counting them to obtain $n_{\mathbf{E},\mathbf{q}}$.

```
partEq <- partE[(partE[, "S"] == "M") & (partE[, "T"] == "car"), ]
nEq <- nrow(partEq)
```

Finally, in step 4 we compute the conditional probability for the query as $n_{\mathbf{E},\mathbf{q}}/n_{\mathbf{E}}$.

```
nEq/nE
[1] 0.34366
```

Algorithm 6.6 Likelihood Weighting Algorithm

---

1. Order the variables in $\mathbf{X}$ according to the topological ordering implied by G, say $X_{(1)} \prec X_{(2)} \prec \ldots \prec X_{(p)}$.

2. Set $w_{\mathbf{E}} = 0$ and $w_{\mathbf{E},\mathbf{q}} = 0$.

3. For a suitably large number of samples $\mathbf{x} = (x_1, \ldots, x_p)$:

   (a) generate $x_{(i)}, i = 1, \ldots, p$ from $X_{(i)} \mid \Pi_{X_{(i)}}$ using the values $e_1, \ldots, e_k$ specified by the hard evidence $\mathbf{E}$ for $X_{i_1}, \ldots, X_{i_k}$;

   (b) compute the weight

   $$w_{\mathbf{x}} = \prod \Pr(X_{i^*} = e_* \mid \Pi_{X_{i^*}});$$

   (c) set $w_{\mathbf{E}} = w_{\mathbf{E}} + w_{\mathbf{x}}$;

   (d) if $\mathbf{x}$ includes $\mathbf{Q} = \mathbf{q}$, set $w_{\mathbf{E},\mathbf{q}} = w_{\mathbf{E},\mathbf{q}} + w_{\mathbf{x}}$.

4. Estimate $\Pr(\mathbf{Q} \mid \mathbf{E}, G, \Theta)$ with $w_{\mathbf{E},\mathbf{q}}/w_{\mathbf{E}}$.

The discrepancy in the last digits is the natural consequence of the noisiness of stochastic simulation. Clearly, such an algorithm can be very inefficient if $\Pr(\mathbf{E})$ is small, because most particles will be discarded without contributing to the estimation of $\Pr(\mathbf{Q} \mid \mathbf{E}, G, \Theta)$. However, its simplicity makes it easy to implement and very general in its application: it allows for very complex specifications of $\mathbf{E}$ and $\mathbf{Q}$ for both $\mathrm{MAP}(\mathbf{Q} \mid \mathbf{E}, \mathcal{B})$ and $\mathrm{CPQ}(\mathbf{Q} \mid \mathbf{E}, \mathcal{B})$.

An improvement over logic sampling, designed to solve this problem, is an application of importance sampling called *likelihood weighting* that is illustrated in Algorithm 6.6. Unlike logic sampling, all the particles generated by likelihood weighting include the evidence $\mathbf{E}$ by design. However, this means that we are not sampling from the original BN any more, but we are sampling from a second BN in which all the nodes $X_{i_1}, \ldots, X_{i_k}$ in $\mathbf{E}$ are fixed. This network is called the *mutilated network*.

```
mutbn <- mutilated(survey.bn, list(E = "high"))
mutbn$E

  Parameters of node E (multinomial distribution)

Conditional probability table:
 high  uni
    1    0
```

As a result, simply sampling from mutbn is not a valid approach. If we do so, the probability we obtain is $\Pr(\mathbf{Q}, \mathbf{E} \mid G, \Theta)$, not the conditional probability $\Pr(\mathbf{Q} \mid \mathbf{E}, G, \Theta)$.

```
particles <- rbn(survey.bn, 10^6)
partQ <- particles[(particles[, "S"] == "M") &
                   (particles[, "T"] == "car"), ]
nQ <- nrow(partQ)
nQ/10^6
 [1] 0.33741
```

The role of the weights is precisely to adjust for the fact that we are sampling from mutbn instead of the original bn. As we can see from step 3(b), the weights are just the likelihood components associated with the nodes of bn we are conditioning on (E in this case) for the particles.

```
w <- logLik(survey.bn, particles, nodes = "E", by.sample = TRUE)
```

Having estimated the weights, we can now perform steps 3(c), 3(d) and 4 and obtain the estimated conditional probability for the query.

```
wEq <- sum(exp(w[(particles[, "S"] == "M") &
               (particles[, "T"] == "car")]))
wE <- sum(exp(w))
wEq/wE
 [1] 0.34275
```

The value of wEq/wE ratio is the same as the exact conditional probability we obtained using **gRain** in Section 1.6.2.1. It is a more precise estimate than that obtained above from logic sampling using the same number of particles.

More conveniently, we can perform likelihood weighting with cpquery by setting method = "lw" and specifying the evidence as a named list with one element for each node we are conditioning on.

```
cpquery(survey.bn, event = (S == "M") & (T == "car"),
        evidence = list(E = "high"), method = "lw")
 [1] 0.34838
```

The estimate we obtain is still very precise despite the fact that we are not increasing the number of particles from the default 500 * nparams.fitted(survey.bn) to n = 10^6 as we did for logic sampling.

At the other end of the spectrum, there are in the literature more complex approximate algorithms that can estimate even very small probabilities with high precision. Two examples are the *adaptive importance sampling* (AIS-BN) scheme by Cheng and Druzdzel (2000) and the *evidence pre-propagation importance sampling* (EPIS-BN) by Yuan and Druzdzel (2003). Both can estimate conditional probabilities as small as 10^{-41}, and they also perform better on large networks. However, their assumptions restrict them to discrete BNs and may require the specification of non-trivial tuning parameters.

6.7 Causal Bayesian Networks

Throughout this book, we have defined BNs in terms of conditional independence relationships and probabilistic properties, without any implication that arcs should represent cause-effect relationships. The existence of equivalence classes of networks indistinguishable from a probabilistic point of view provides a simple proof that arc directions are not indicative of causal effects.

However, from an intuitive point of view it can be argued that a "good" BN should represent the causal structure of the data it is describing. Such BNs are usually fairly sparse, and their interpretation is at the same time clear and meaningful, as explained by Pearl (2009) in his book on causality:

> "It seems that if conditional independence judgments are byproducts of stored causal relationships, then tapping and representing those relationships directly would be a more natural and more reliable way of expressing what we know or believe about the world. This is indeed the philosophy behind causal BNs."

This is the reason why building a BN from expert knowledge in practice codifies known and expected causal relationships for a given phenomenon.

Learning such causal models, especially from observational data, presents significant challenges. In particular, three additional assumptions are needed:

- Each variable $X_i \in \mathbf{X}$ is conditionally independent of its non-effects, both direct and indirect, given its direct causes. This assumption is called the *causal Markov assumption*, and represents the causal interpretation of the local Markov property in Definition 6.4.

- There must exist a DAG which is faithful to the probability distribution $\mathbf{P}$ of $\mathbf{X}$, so that the only dependencies in $\mathbf{P}$ are those arising from d-separation in the DAG.

- There must be no *latent variables* (unobserved variables influencing the variables in the network) acting as *confounding factors*. Such variables may induce spurious correlations between the observed variables, thus introducing bias in the causal network. Even though this is often listed as a separate assumption, it is really a corollary of the first two: the presence of unobserved variables violates the faithfulness assumption (because the network structure cannot include them) and possibly the causal Markov property (if an arc is wrongly added between the observed variables due to the influence of the latent one). Note that including those latent variables in the BN as nodes which take value NA for all observations, connecting them with observed nodes and estimating their values in the process of learning the BN can take care of this issue by letting the BN take their effect into account.

These assumptions are difficult to verify in real-world settings, as the set of the potential confounding factors is not usually known. At best, we can address this issue, along with selection bias, by implementing a carefully planned experimental design.

Furthermore, even when dealing with interventional data collected from a scientific experiment (where we can control at least some variables and observe the resulting changes), there are usually multiple equivalent BNs that represent reasonable causal models. Many arcs may not have a definite direction, resulting in substantially different DAGs. When the sample size is small there may also be several non-equivalent BNs fitting the data equally well. Therefore, in general we are not able to identify a single, "best", causal BN but rather a small set of likely causal BNs that fit our knowledge of the data.

An example of the bias introduced by the presence of a latent variable was illustrated by Edwards (2000, page 113) using the marks data. marks, which is included in the **bnlearn** package, consists of the exam scores of 88 students across five different topics, namely: mechanics (MECH), vectors (VECT), algebra (ALG), analysis (ANL) and statistics (STAT). The scores are bounded in the interval $[0, 100]$. This data set was originally investigated by Mardia et al. (1979) and subsequently in classical books on graphical models such as Whittaker (1990).

```
data(marks)
head(marks, n = 5)
   MECH VECT ALG ANL STAT
1    77   82  67  67   81
2    63   78  80  70   81
3    75   73  71  66   81
4    55   72  63  70   68
5    63   63  65  70   63
```

Edwards noted that the students apparently belonged to two groups (which we will call A and B) with substantially different academic profiles. He then assigned each student to one of those two groups using the Expectation-Maximisation (EM) algorithm to impute group membership as a latent variable (LAT). The EM algorithm assigned the first 52 students (with the exception of number 45) to belong to group A, and the remainder to group B.

```
latent <- factor(c(rep("A", 44), "B", rep("A", 7), rep("B", 36)))
modelstring(hc(marks[latent == "A", ]))
 [1] "[MECH][ALG|MECH][VECT|ALG][ANL|ALG][STAT|ALG:ANL]"
modelstring(hc(marks[latent == "B", ]))
 [1] "[MECH][ALG][ANL][STAT][VECT|MECH]"
modelstring(hc(marks))
 [1] "[MECH][VECT|MECH][ALG|MECH:VECT][ANL|ALG][STAT|ALG:ANL]"
```

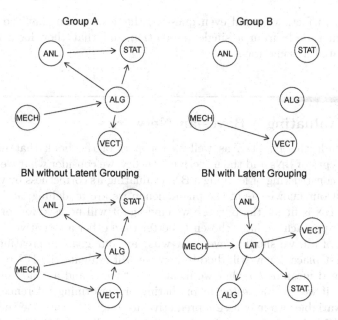

Figure 6.7
BNs learned from the marks data set when considering only group A (top left), when considering only group B (top right), when considering both (bottom left) and from discretised data set after the inclusion of the latent variable LAT (bottom right).

As we can see from Figure 6.7 (top-left and top-right panels) and from the modelstring calls above, the BNs learned from group A and from group B are completely different. Furthermore, they are both different from the BN learned from the whole data set (bottom-left panel).

If we want to learn a single BN while taking LAT into account, we can either use a CGBN (as we will do in Chapter 7) or discretise the marks data and include the latent variable when learning the structure of the (now multinomial) BN. Again, we obtain a BN whose DAG (bottom-right panel) is completely different from those above.

```
dmarks <- discretize(marks, breaks = 2, method = "interval")
modelstring(hc(cbind(dmarks, LAT = latent)))
 [1] "[MECH][ANL][LAT|MECH:ANL][VECT|LAT][ALG|LAT][STAT|LAT]"
```

This BN provides a simple interpretation of the relationships between the topics: the grades in mechanics and analysis can be used to infer which group a student belongs to, and that in turn influences the grades in the remaining topics. We can clearly see that any causal relationship we would have inferred from a DAG learned without taking LAT into account would be potentially spurious. In fact, we could even question the assumption that the data are a random sample from a single population and that they have not been manipulated in some way.

6.8 Evaluating a Bayesian Network

To conclude this chapter, as well as the part of the book that introduces various types of BNs and the underlying theory, we consider what we ought to do when constructing or learning a BN: evaluating its correctness by validating it against our knowledge of the phenomenon we are modelling; and checking that the BN is fit for the purpose we envisage it will be used for, or that the learning approach we have chosen is better than other alternatives.

First of all, we should have clear what are our *goals* in creating the BN in the first place. They will dictate how we will evaluate the BN by giving a well-defined meaning to what we mean by a "good" and a "bad" model. For instance, if we are interested in predicting or performing inference on some specific variables we may assess predictive accuracy. We saw one such case in Section 4.5, when we were trying to predict air quality 10 or 20 minutes in the future with a DBN. For static data, we can use cross-validation to estimate predictive accuracy in different ways. For instance, the code below runs 10-fold cross-validation for 10 times, learns a BN from the training folds, and computes the negated log-likelihood of the validation folds for all variables.

```
bn.cv(marks, bn = "hc", method = "k-fold", k = 10, runs = 10)

  k-fold cross-validation for Bayesian networks

  target learning algorithm:              Hill-Climbing
  number of folds:                        10
  loss function:
                                          Log-Likelihood Loss (Gauss.)
  number of runs:                         10
  average loss over the runs:             19.498
  standard deviation of the loss:         0.022816
```

This provides an empirical approximation of the Kullback-Leibler distance between the BN we learn and the unobservable "true" BN; hence lower values are better. The intuition behind this loss is that, if the BN is an accurate description of the phenomenon we are modelling and if the new data have the same distribution as the data the BN was learned from, the log-likelihood of BN for the new data should be high (and the negated log-likelihood should then be low).

Another example, in which we focus specifically on predicting a single variable (algebra marks), is the following.

```
bn.cv(marks, bn = "hc", method = "k-fold", k = 10, runs = 10,
  loss = "cor-lw", loss.args = list(target = "ALG"))

  k-fold cross-validation for Bayesian networks

  target learning algorithm:          Hill-Climbing
  number of folds:                    10
  loss function:
                      Predictive Correlation (Posterior, Gauss.)
  training node:                      ALG
  number of runs:                     10
  average loss over the runs:         0.79111
  standard deviation of the loss:     0.0097304
```

In this case loss is the correlation between the values of ALG in the validation folds and those predicted by the BNs learned from the training folds; hence higher values are better. Equivalently we could use the mean square error as a loss with loss = "mse-lw": one is inversely proportional to the other, so they provide the same information just on different scales. The corresponding choice for discrete data is loss = "pred-lw", the classification error. The "-lw" suffix indicates that the values of the loss functions are computed using posterior predictions obtained by likelihood weighting.

Instead of evaluating a learning strategy (hill-climbing above, hence bn = "hc"), we can also validate a specific network structure by passing it to bn.cv via the bn argument.

```
dag.check <- model2network("[MECH][ALG|MECH][VECT][ANL|ALG][STAT|ANL]")
bn.cv(marks, bn = dag.check, method = "k-fold", k = 10, runs = 10,
  loss = "cor-lw", loss.args = list(target = "ALG"))

  k-fold cross-validation for Bayesian networks

  target network structure:
   [MECH][VECT][ALG|MECH][ANL|ALG][STAT|ANL]
  number of folds:                    10
  loss function:
                      Predictive Correlation (Posterior, Gauss.)
  training node:                      ALG
  number of runs:                     10
  average loss over the runs:         0.7345
  standard deviation of the loss:     0.0063171
```

If, on the other hand, our aim is to construct a mechanistic model of reality, we may take a causal perspective and look for the presence or absence

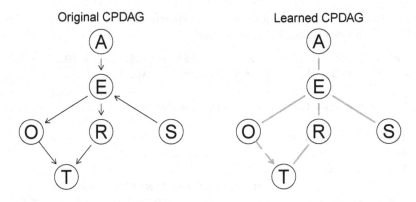

Figure 6.8
CPDAG comparison using `graphviz.compare`. False positive and false negative
arcs are in grey; false negative arcs are dashed lines.

of particular (patterns of) arcs. In that case, we may also collect information
available from experts that was not used in constructing the model and check
whether the BN agrees with it. One example will be shown in Section 8.1,
where the presence of particular biological reactions in a BN learned from data
was validated using reports confirming their existence. A simpler example,
using the `survey` example from Chapter 1, would be checking the Structural
Hamming Distance (SHD) between the BN we learned from the data and that
we constructed manually.

```
shd(survey.dag, hc(survey))
 [1] 6
```

SHD measures the number of arcs that differ between the CPDAG of the first
BN and that of the second, thus considering only those arc directions that
can be uniquely identified. The `compare` function can give us more detailed
information about what those differences are in terms of false positives ("fp")
and false negatives ("fn").

```
diff <- compare(cpdag(survey.dag), cpdag(hc(survey)))
unlist(diff)
 tp fp fn
  0  5  6
```

We can see that there are no true positives ("tp"), so the two CPDAGs share
no arcs: 6 are missing, and 5 appear where they are not supposed to. Adding
`arcs = TRUE` to `compare` will return which arcs are in `tp`, `fp` and `fn` instead of
the counts for each group. Another option is to compare these two CPDAGs
visually with `graphviz.compare`, which produces the plot in Figure 6.8.

```
graphviz.compare(cpdag(survey.dag), cpdag(hc(survey)),
   main = c("Original CPDAG", "Learned CPDAG"),
```

```
diff.args = list(fn.col = "grey", fn.lwd = 4, fn.lty = "longdash",
                 fp.col = "grey", fp.lwd = 4))
```

In addition to the layout options available from the simpler graphviz.plot, graphviz.compare takes care of placing the nodes in the same position in all the panels to make the plot easy to read; and it internally calls compare to find the differences between the bn objects it gets as arguments.

Another issue that should be considered is the *stability* of the BN, that is, how sensitive it is to small changes in the data and in the tuning parameters of the methods used to construct it. A common way to assess that is to perturb them and checking that small changes in the data or in the tuning parameters result in small changes in the BN. In the context of structure learning, boot.strength uses bootstrap resampling to perturb the data and quantify the probability that two nodes are connected (the strength column) and the probability of a particular arc direction (the direction column) by learning multiple DAGs.

```
head(boot.strength(survey, algorithm = "hc"), n = 4)
   from to strength direction
1    A  R    0.290   0.50862
2    A  E    0.665   0.53008
3    A  O    0.115   0.50000
4    A  S    0.005   0.50000
```

If structure learning is stable, we expect to see a few arcs with high probabilities (ideally those we learned) and most with probabilities close to zero. If that is the case then our structure learning approach can reliably discriminate between well-supported and ill-supported arcs. The probabilities above can also be used to construct a consensus DAG containing only arcs with strong support from the data, as we will discuss in detail in Section 8.1.3.

A more general take on the same approach is implemented in the bn.boot function, which will measure an arbitrary function of the DAGs. A simple example would be the number of arcs in the DAG, which can be computed using the narcs function.

```
boot.stats <- bn.boot(survey, statistic = narcs, algorithm = "hc",
                algorithm.args = list(score = "bde", iss = 1))
summary(unlist(boot.stats))
   Min. 1st Qu.  Median    Mean 3rd Qu.    Max.
   2.00    4.00    4.00    4.42    5.00    9.00
boot.stats <- bn.boot(survey, statistic = narcs, algorithm = "hc",
                algorithm.args = list(score = "bde", iss = 5))
summary(unlist(boot.stats))
   Min. 1st Qu.  Median    Mean 3rd Qu.    Max.
   2.00    5.00    6.00    5.96    7.00   10.00
```

From the output above, we can see that changing the imaginary sample size of BDe from 1 to 5 increases the expected number of arcs from about 4.5 to

6. But we can also see that while 75% of the arc sets contain $[4, 7]$ arcs for iss = 1, 75% contain $[5, 10]$ arcs for iss = 5 suggesting more unstable learned DAGs. The imaginary sample size indeed can influence structure learning in various ways depending on the distribution of the data: its behaviour is formally characterised in Ueno (2011).

The stability of the estimates of the parameters of the BN for a given structures can be assessed using bootstrap in a similar way.

6.9   Further Reading

Basic definitions and properties of BNs are detailed in many books, each with its own different perspective; but perhaps the most clear and concise are still the seminal works Pearl (1988) and Pearl (2009). We would also like to suggest Castillo et al. (1997) for a formal introduction to the theory of graphical models; and the tome from Koller and Friedman (2009), which is a very comprehensive if lengthy reference. Different aspects are also covered in Murphy (2012).

Various inference approaches are covered in different books: from variable elimination in Koller and Friedman (2009, Chapter 9) and Russell and Norvig (2009, Section 14.4); to junction trees in Koller and Friedman (2009, Chapter 10), Korb and Nicholson (2011, Chapter 3), Castillo et al. (1997, Chapter 8) and Koski and Noble (2009, Chapter 10); to logic sampling and likelihood weighting in Koller and Friedman (2009, Chapter 12), Korb and Nicholson (2011, Chapter 3) and Castillo et al. (1997, Chapter 9). Pearl (1988, Chapters 4 and 5) is also an interesting read about exact inference. In addition, Koller and Friedman (2009, Chapter 13) describes more inference algorithms specific to MAP queries.

Learning has fewer references compared to inference, and structure learning even fewer still. Parameter learning for discrete BNs is covered in most books, including Koller and Friedman (2009, Chapter 17) and Neapolitan (2003, Section 7.1); GBNs are covered in Neapolitan (2003, Section 7.2) and in Koller and Friedman (2009, Section 17.2.4). Constraint-based structure learning is introduced in Korb and Nicholson (2011, Chapter 8) and Neapolitan (2003, Chapter 10); Edwards (2000, Chapter 5) provides an exhaustive list of conditional independence tests to use in this context. Score-based structure learning is introduced in Korb and Nicholson (2011, Chapter 9) and Castillo et al. (1997, Chapter 11), and a more general take on the algorithms that can be used for this task is in Russell and Norvig (2009, Chapter 4). Hybrid structure learning is mentioned in Koski and Noble (2009, Section 6.3).

Finally, for the interested reader, we suggest Spirtes et al. (2000) as a very thorough reference to causal BNs and using constraint-based algorithms such as PC to learn them.

Exercises

Exercise 6.1 *Consider the network below.*

```
dag <- model2network("[A][C|A:B][D|C:A][F|A][B][E|B:D:F]")
```

1. *What are the minimal set(s) of variables required to d-separate C and E (that is, sets of variables for which no proper subset d-separates C and E)?*

2. *What are the minimal set(s) of variables required to d-separate A and B?*

3. *What are the maximal set(s) of variables that d-separate C and E (that is, sets of variables for which no proper superset d-separates C and E)?*

4. *What are the maximal set(s) of variables that d-separate A and B?*

5. *Using the chain rule, establish an expression for the joint distribution over {A, B, C, D, E, F}. Use this expression to show that B and D are conditionally independent given A and C.*

Exercise 6.2 *Let A be a node in a DAG. Assume that all variables in the Markov blanket of A are instantiated. Show that A is d-separated from the remaining uninstantiated variables.*

Exercise 6.3 *Consider the following DAGs.*

```
dag1 <- model2network("[G][H][D|G][I|G:H][F|H][B|D][E|G][C|E:F][A|B:C]")
dag2 <-
  model2network("[B][C][A][D|B][E|B][F|C][G|A:D][H|D:E][I|E:F][J|G]")
```

1. *Determine which variables are d-separated from A given B (in dag1) and which variables are d-separated from A given J (in dag2).*

2. *Apply the procedure shown in the Section 6.4 to determine whether A is d-separated from F given {H, I} in dag1.*

3. *Similarly, determine whether A is d-separated from C given B in dag2.*

Exercise 6.4 *Consider the* survey *data set from Chapter 1.*

1. Learn a BN with the IAMB algorithm and the asymptotic mutual information test.

2. Learn a second BN with IAMB but using only the first 100 observations of the data set. Is there a significant loss of information in the resulting BN compared to the BN learned from the whole data set?

3. Repeat the structure learning in the previous point with IAMB and the Monte Carlo and sequential Monte Carlo mutual information tests. How do the resulting networks compare with the BN learned with the asymptotic test? Is the increased execution time justified?

Exercise 6.5 *Consider again the* survey *data set from Chapter 1.*

1. Learn a BN using Bayesian posteriors for both structure and parameter learning, in both cases with iss = 5.

2. Repeat structure learning with hc *and 3 random restarts and with* tabu. *How do the BNs differ? Is there any evidence of numerical or convergence problems?*

3. Use increasingly large subsets of the survey *data to check empirically that BIC and BDe are asymptotically equivalent.*

Exercise 6.6 *Consider the* marks *data set from Section 6.7.*

1. Create a bn *object describing the graph in the bottom-right panel of Figure 6.7 and call it* mdag.

2. Construct the skeleton, the CPDAG and the moral graph of mdag.

3. Discretise the marks *data using* "interval" *discretisation with 2, 3 and 4 intervals.*

4. Perform structure learning with hc *on each of the discretised data sets; how do the resulting DAGs differ?*

7

Software for Bayesian Networks

The difficulty of implementing versatile software for general classes of graphical models and the varying focus in different disciplines limit the applications of BNs compared to the state of the art in the literature. Nevertheless, the number of R packages for BNs has been slowly increasing in recent years. In this chapter we will provide an overview of available software packages, without pretending to be exhaustive, and we will introduce some classic R packages dealing with different aspects of BN learning and inference.

7.1 An Overview of R Packages

There are several packages on CRAN dealing with BNs. They can be divided in two categories: those that implement structure and parameter learning and those that focus only on parameter learning and inference (see Table 7.1).

Packages **bnlearn** (Scutari, 2010), **deal** (Bøttcher and Dethlefsen, 2003), **pcalg** (Kalisch et al., 2012), **catnet** (Balov and Salzman, 2020) and **abn** (Kratzer et al., 2019) fall into the first category.

bnlearn offers a wide variety of structure learning algorithms (spanning all the three classes covered in this book, with several tests and network scores), parameter learning approaches (maximum likelihood for discrete and continuous data, Bayesian estimation for discrete data) and inference techniques (cross-validation, bootstrap, conditional probability queries and prediction). It is also the only package that keeps a clear separation between the structure of a BN and the associated local probability distributions; they are implemented as two different classes of R objects.

deal implements structure and parameter learning using a Bayesian approach. It is one of the few R packages that handles BNs combining discrete and continuous nodes with CGBNS. The network structure is learned with the hill-climbing greedy search described in Algorithm 6.2, with the posterior density of the network as a score function and random restarts to avoid local maxima.

pcalg provides a free software implementation of the PC algorithm, and it is specifically designed to estimate and measure causal effects. It handles both discrete and continuous data, and can account for the effects of latent variables

	bnlearn	catnet	deal	pcalg
discrete data	Yes	Yes	Yes	Yes
continuous data	Yes	No	Yes	Yes
mixed data	No	No	Yes	No
constraint-based learning	Yes	No	No	Yes
score-based learning	Yes	Yes	Yes	No
hybrid learning	Yes	No	No	No
structure manipulation	Yes	Yes	No	No
parameter estimation	Yes	Yes	Yes	Yes
prediction	Yes	Yes	No	No
inference	Yes	No	No	No

	abn	gRbase	gRain	rbmn
discrete data	Yes	Yes	Yes	No
continuous data	Yes	Yes	No	Yes
mixed data	Yes	No	No	No
constraint-based learning	No	No	No	No
score-based learning	Yes	No	No	No
hybrid learning	No	No	No	No
structure manipulation	Yes	Yes	No	No
parameter estimation	Yes	No	No	Yes
prediction	No	No	Yes	Yes
inference	No	No	Yes	No

Table 7.1
Feature matrix for the R packages covered in Section 7.1.

on the network. The latter is achieved through a modified PC algorithm known as *Fast Causal Inference* (FCI), first proposed by Spirtes et al. (2000).

abn implements *additive BNs* that model the local distribution of each node with a generalised linear model, thus allowing a greater degree of flexibility compared to CGBNs. This package, along with the companion **mcmcabn** package (Kratzer and Furrer, 2019), focuses on structure learning from a fully Bayesian perspective while providing some support for simulations and model validation.

catnet models discrete BNs using frequentist techniques. Structure learning is performed in two steps. First, the node ordering of the DAG is learned from the data using simulated annealing; as an alternative, a custom node ordering can be specified by the user. An exhaustive search is then performed among the network structures compatible with the given node ordering and the exact maximum likelihood solution is returned. Parameter learning and prediction are also implemented.

Packages **gRbase** (Højsgaard et al., 2020) and **gRain** (Højsgaard, 2020) fall into the second category. They focus on manipulating the parameters of the network, on prediction and on inference, under the assumption that all variables are discrete. Neither **gRbase** nor **gRain** implement any structure or parameter learning algorithm, so the BN must be completely specified by the user.

rbmn (Denis, 2020) is devoted to linear GBNs and more specifically on deriving closed form expressions for the joint and conditional distributions of subsets of nodes. Conversion functions to and from **bnlearn** objects are available. No structure learning is implemented.

7.1.1 The deal Package

The **deal** package implements structure learning following step-by-step a classic Bayesian workflow, which we will illustrate using the marks data set from Section 6.7. First of all, we load the marks data set from **bnlearn** and we add the latent grouping, so that we have both discrete and continuous variables in the data.

```
library(bnlearn)
data(marks)
latent <- factor(c(rep("A", 44), "B", rep("A", 7), rep("B", 36)))
latent.marks <- data.frame(marks, LAT = latent)
```

The first step consists in defining an object of class network which identifies the variables in the BN and whether they are continuous or discrete.

```
library(deal)
net <- network(latent.marks)
```

```
net
## 6 ( 1 discrete+ 5 ) nodes;score=  ;relscore=
1    MECH    continuous()
2    VECT    continuous()
3    ALG     continuous()
4    ANL     continuous()
5    STAT    continuous()
6    LAT     discrete(2)
```

Subsequently, we define an uninformative prior distribution on net with jointprior and we set the imaginary sample size to 5.

```
prior <- jointprior(net, N = 5)
Imaginary sample size: 5
```

Once we have defined the prior, we can perform a hill-climbing search for the DAG with the highest posterior probability using autosearch.

```
net <- learn(net, latent.marks, prior)$nw
best <- autosearch(net, latent.marks, prior)
```

The posterior probability is a combination of BDe and BGe, since we are learning a CGBN. In particular:

- Discrete nodes can only have other discrete nodes as their parents, and their contributions to the posterior probability are the same as in the BDe score.

- Continuous nodes can have both continuous and discrete parents. The part of the posterior probability that derives from continuous parents has the same form as the BGe score, while the part that derives from discrete parents is constructed like a mixture.

It is important to note that **deal** has a modelstring function which produces a compact string representation of a DAG in the same format as **bnlearn**'s modelstring. This is very convenient to import and export DAGs between the two packages.[1]

```
mstring <- deal::modelstring(best$nw)
bnlearn:::fcat(mstring)
   [MECH|VECT:ALG][VECT|LAT][ALG|VECT:LAT][ANL|ALG:LAT][STAT|ALG:ANL]
   [LAT]
```

As we can see in the output above, LAT is a root node, because it is the only discrete node and other nodes can only be its descendants. It is also important to note that since **deal** and **bnlearn** provide many functions with identical names, using the explicit :: notation is advisable.

---

[1] fcat is an internal function in **bnlearn** that is used by print to wrap long model strings. It is not exported and may change in the future.

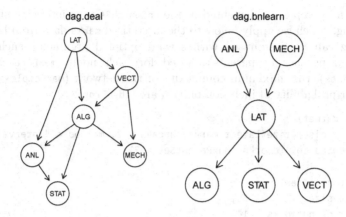

Figure 7.1
DAGs learned from the marks data with **deal** (left) and **bnlearn** (right) after
adding latent grouping.

Using compare we can see that the DAG we just learned (dag.deal) is
completely different from that in Section 6.7 (here dag.bnlearn).

```
dag.deal <- model2network(mstring)
unlist(bnlearn::compare(cpdag(dag.deal), cpdag(dag.bnlearn)))
 tp fp fn
  0  5  9
```

The two DAGs, which are shown in Figure 7.1, have CPDAGs that do not
share any arc! In fact, dag.deal is quite similar to the GBN in the bottom-
left panel of Figure 6.7, which was learned before introducing LAT. This is,
once more, an example of how different parametric assumptions and data
transformation can deeply affect the results of BN learning.

7.1.2 The catnet Package

The **catnet** package also divides structure learning in discrete steps, but from
a frequentist perspective: it learns the DAG that maximises BIC instead of
posterior probability. Following Friedman and Koller (2003), **catnet** starts
from the consideration that finding the optimal DAG given a topological
ordering is a much simpler task than doing the same in the general case.
Therefore, it implements the following workflow:

1. learn the optimal topological ordering from the data, or let the user
   specify one; and then

2. learn the optimal structure of the DAG conditional on the topological
   ordering.

The first step is implemented in the cnSearchSA function using simulated annealing, which we apply below to the discretised marks data from **bnlearn**. Learning can be customised with several optional arguments such as the maximum number of parents allowed for each node (maxParentSet and parentSizes), the maximum complexity of the network (maxComplexity) and the prior probability of inclusion of each arc (edgeProb).

```
library(catnet)
dmarks <- discretize(latent.marks, breaks = 2, method = "interval")
ord <- cnSearchSA(dmarks, maxParentSet = 2)
ord
  Number of nodes    = 6,
  Sample size        = 88,
  Number of networks = 14
  Processing time    = 0.114
```

ord is an object of class catNetworkEvaluate, which contains a set of networks compatible with the learned topological ordering, along with other useful quantities. We can access them as follows, and we can explore their properties with various **catnet** functions.

```
nets <- ord@nets
nets[[1]]
  A catNetwork object with  6  nodes,  0  parents,  2  categories,
   Likelihood =  -3.89 , Complexity =  6 .
```

The second step is implemented in cnFindBIC, which finds the network with the best BIC score among those stored in ord.

```
best <- cnFindBIC(ord, nrow(dmarks))
best
  A catNetwork object with  6  nodes,  1  parents,  2  categories,
   Likelihood =  -2.96 , Complexity =  11 .
```

Unlike **deal**, **catnet** provides functions to explore the learning process and the resulting networks and to make use of the latter. For instance, we can use cnSamples to generate numsamples random samples from best, which is useful to implement approximate inference. They are returned in a data frame with the same structure as that used to learn best.

```
cnSamples(best, numsamples = 4)
        MECH        VECT        ALG         ANL       STAT LAT
1 (38.5,77]  [9,45.5]  [15,47.5]  [9,39.5]   [9,45]    B
2  [0,38.5] (45.5,82] (47.5,80] (39.5,70]  (45,81]    A
3  [0,38.5] (45.5,82] (47.5,80] (39.5,70]   [9,45]    A
4  [0,38.5] (45.5,82] (47.5,80] (39.5,70]   [9,45]    A
```

Another possibility is to extract the arc set from best using cnMatEdges, and import the DAG in **bnlearn**.

```
em <- empty.graph(names(dmarks))
arcs(em) <- cnMatEdges(best)
```

This makes it possible to use other functions in **bnlearn**, such as bn.fit to learn the parameters, and to export the BN to **gRain** to perform exact inference.

7.1.3   The pcalg Package

The **pcalg** package is unique in its focus on causal inference. Unlike other packages, it allows the user to provide a custom function to perform conditional independence tests. This means that, in principle, we can model any kind of data as long as we can define a suitable test function. gaussCItest (Student's t test for correlation), condIndFisherZ (Fisher's Z test), gSquareDis (G^2 test) and discItest (Pearson's X^2 test) implement commonly used tests, and make the analysis of discrete BNs and GBNs possible out of the box.

However, this flexibility requires more effort in setting up structure learning: the user must produce the sufficient statistics for the conditional independence tests from the data as a preliminary step. In the case of the marks data (minus the latent grouping), this means computing the sample size and the correlation matrix of the variables.

```
library(pcalg)
suffStat <- list(C = cor(marks), n = nrow(marks))
```

We can then call the pc function to use the PC algorithm with the specified test (indepTest) and type I error threshold (alpha) and learn the DAG.

```
pc.fit <- pc(suffStat, indepTest = gaussCItest,
             labels = colnames(marks), alpha = 0.05)
```

Note that the data are never used directly, but only through the sufficient statistics in suffStat. The pc.fit object stores the DAG in an object of class graphNEL, which is defined in the **graph** package.

```
pc.fit@graph
A graphNEL graph with directed edges
Number of Nodes = 5
Number of Edges = 7
```

As a result, it is easy to import the DAG in **bnlearn** with the as.bn function, or to manipulate and explore it directly with **graph** and **Rgraphviz**.

The FCI algorithm, however, is more appropriate for the marks data because of the influence of the latent variable LAT. The syntax of the fci function is the same as that of pc, differing only in some optional arguments which are not used here.

```
fci.fit <- fci(suffStat, indepTest = gaussCItest,
               labels = colnames(marks), alpha = 0.05)
```

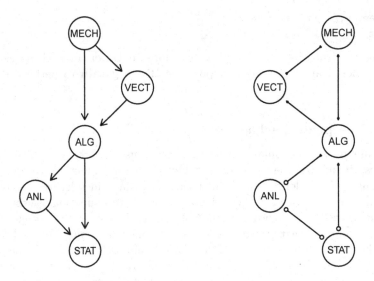

Figure 7.2
Graphs learned from the marks data with the PC (left) and FCI (right)
algorithms without considering latent grouping.

A visual comparison of pc.fit and fci.fit, shown in Figure 7.2, makes it
clear the two graphs should be interpreted differently even though they look
very similar. fci.fit has several types of arrowheads, which are not present
in pc.fit, to distinguish between the causal effects originating for observed
variables as opposed to latent ones. A detailed treatment of their meaning,
which would be quite involved, is beyond the scope of this overview; we refer
the interested reader to **pcalg**'s documentation and the literature referenced
therein.

7.1.4 The abn Package

The **abn** package's key features are the wide choice of distribution for the local
distributions (Gaussian, Poisson, multinomial, binomial), its focus on Bayesian
structure and parameter learning, and the availability of both exact and
heuristic structure learning algorithms. Each node is modelled as a generalised
linear model:

$$E(X_i \mid \Pi_{X_i}) = g^{-1}\left(\Pi_{X_i}\beta_{X_i} + \varepsilon_{X_i}\right),$$

so that each variable X_i is explained by a linear combination of its parents
Π_{X_i} with regression coefficients β_{X_i}. ε_{X_i} is the error term. The function $g(\cdot)$
is called the *link function* and its purpose is to transform the linear predictor
$\Pi_{X_i}\beta_{X_i}$ into the right scale for distribution we have assumed for the node. For

instance, if we assume that X_i follows a normal distribution, **abn** sets $g(\cdot)$ to the identity function to obtain a linear regression model $X_i = \Pi_{X_i}\beta_{X_i} + \varepsilon_{X_i}$ much like in Chapter 2; if we assume that X_i follows a binomial distribution, **abn** sets $g(\cdot) = \log(p_{X_i}/(1-p_{X_i}))$ to obtain a logistic regression; if we assume that X_i follows a Poisson distribution, **abn** sets $g(\cdot) = \log(\lambda_{X_i})$ to obtain a log-linear regression.

Going back to the latent.marks one last time, we can put this flexibility to use by modelling the marks as Poisson random variables; which is arguably a more fitting distributional assumption than normality since the marks are positive integers. In order to do so, we enter the distribution assumed for each variable in a list, which we name dists below.

```
library(abn)
dists <- list(MECH = "poisson", VECT = "poisson", ALG = "poisson",
              ANL = "poisson", STAT = "poisson", LAT = "binomial")
```

With this information, buildscorecache can use the data (latent.marks in this example) to create a cache of the score components arising from the local distributions following Equation (6.14).

```
cache <- buildscorecache(latent.marks, data.dists = dists,
             dag.banned = ~ LAT | .)
```

The dag.banned argument can be used to blacklist arcs and reduce the number of score components that are computed from the data. In the above, we

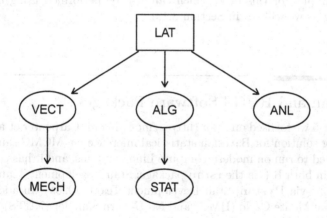

Figure 7.3
DAG learned with the **abn** package from the marks data when considering latent grouping and assuming marks are distributed as Poisson random variables. Discrete variables are framed with a rectangle, continuous variables with an ellipse.

blacklist all the arcs pointing to the LAT node to obtain a DAG comparable to that in Section 6.7.

At this point we can choose whether to use exact or heuristic structure learning algorithms. mostprobable implements the exact algorithm Koivisto and Sood (2004), which searches over the topological ordering of the nodes as well as the DAGs to speed up learning. searchHillclimber implements hill climbing, which we run with 20 random restarts using the num.searches argument.

```
exact.net <- mostprobable(cache, verbose = FALSE)
hc.net <- searchHillclimber(cache, num.searches = 20)
```

The two DAGs exact.net and hc.net are identical and are shown in Figure 7.3 (generated with the plotabn function): we can see strong similarities with dag.deal and some with that in Figure 6.7.

The parameters can be learned with fitabn, which defaults to Bayesian posterior estimates.

```
fitted <- fitabn(exact.net)
```

The score cache created by buildscorecache can also be passed to the mcmcabn function in the **mcmcabn** package to perform structural inference: DAGs will be sampled from $\Pr(G \mid \mathcal{D})$ using MCMC resampling. From these DAGs we can compute the posterior probabilities of particular structural features, such as arcs or sub-networks, and assess the stability of the learned network. The **BiDAG** (Kuipers and Moffa, 2017) package implements another MCMC sampler for this task, which can also be performed using bootstrap resampling as we will see in Section 8.1.3.

7.2 Stan and BUGS Software Packages

In Chapter 5 we focused on Stan (http://mc-stan.org; Carpenter et al., 2017), the leading solution for Bayesian statistical inference via MCMC sampling. It is well suited to run on modern compute Linux systems, and it has convenient interfaces in both R (via the **rstan** package; Stan Development Team, 2020b) and Python (via **Pystan**; Stan Development Team, 2020a). In addition, its Hamiltonian Monte Carlo (HMC) and No-U-Turn Sampler (NUTS) sampling algorithms are efficient and numerically stable.

Stan is the latest in a long lineage of software packages implementing MCMC samplers, starting from various BUGS (Bayesian Inference Using Gibbs Sampling) implementations such as OpenBUGS (Lunn et al., 2009) and WinBUGS (Lunn et al., 2000) to JAGS (Just Another Gibbs Sampler; Plummer, 2003). As their names suggest, they are all based on Gibbs sampling. For models with a large number of parameters, models with non-conjugate prior distributions, and models whose parameters have strongly correlated posterior distributions, Stan is generally faster at converging to and sampling from the correct posterior distribution.

7.2.1 Stan: A Feature Overview

Stan has several features that make it the best options for MCMC Bayesian inference. First of all, it has built-in support for a wide range of probability distributions including:

- all common (and some not-so-common) discrete and continuous distributions;

- conditional distributions, which in fact implement regression models such as logistic regression, negative binomial Regression, Poisson regression and Gaussian dynamic linear models;

- specialised distribution for extreme values analysis (such as the Gumbel distribution) and for reliability/survival analysis (such as Weibull distribution).

In addition, users can create custom distributions directly within Stan, specifying parameter types and their ranges. This flexibility is made possible by advanced architectural features like symbolic automatic differentiation (Carpenter et al., 2015) and automatic variational inference (Kucukelbir et al., 2015).

Secondly, users can also implement custom functions directly within Stan, including functions that can access and affect the state of the sampler (that is, perform sampling themselves) and recursive functions. Stan itself, however, provides a comprehensive selection of mathematical functions, including composed functions to achieve better numeric precision (like $\texttt{log1m(x)} = \log(1-x)$ and $\texttt{expm1(x)} = e^{x-1}$.); array, matrix and sparse matrix operations; equation and ordinary differential equation solvers.

Finally, and unlike its predecessors, Stan is highly embeddable and easy to tune for different hardware because it translates models to C++ code under the hood. In R we can do this explicitly with the $\texttt{stanc}$ command of **rstan**.

```
library(rstan)
model <- 'data {
  int <lower=0> N;
  int <lower=0,upper=1> y[N];
}
parameters {
  real <lower=0,upper=1> theta;
}
model {
```

```
  theta ~ beta(1,1);
  y ~ bernoulli(theta);
}'
cc <- stanc(model_code = model, model_name = "beta-binomial")
cppcode <- unlist(strsplit(cc$cppcode, "\\n"))
cat(head(cppcode, n = 15), "...", sep = "\n")
// Code generated by Stan version 2.21.0

#include <stan/model/model_header.hpp>

namespace modelb2d85e2561ef_beta_binomial_namespace {

using std::istream;
using std::string;
using std::stringstream;
using std::vector;
using stan::io::dump;
using stan::math::lgamma;
using stan::model::prob_grad;
using namespace stan::math;

...
```

Hence, raw performance is limited only by how well can the system's C++ compiler can optimise code (vectorising, using intrinsics, parallelising, etc.).

7.2.2 Inference Based on MCMC Sampling

MCMC algorithms are not straightforward to use for inference, as their output can easily be affected by a number of statistical and numerical issues. The danger is that we have no way of checking whether the numerical results we obtain are valid or not. It is beyond the scope of this book to cover this topic in much detail: we refer interested readers to Robert and Casella (2009). A few, important points are sketched below.

MCMC samples are drawn from the joint posterior distribution of the parameters or of the variables of interest. They are obtained with complex and complicated iterative algorithms in which the values drawn in the ith iteration depend on those drawn it the $(i-1)$th iteration. Therefore:

• Successive draws are not independent, especially when the algorithm is said to be *not mixing well*. In that case many iterations are needed for some nodes to explore the range of their possible values and generate markedly different values. In other words, samples are autocorrelated. Fortunately, autocorrelation can be easily detected and weakened by keeping one sample every k for a suitably large k. This process is called *thinning*.

• It can be shown that MCMC algorithms sample from the desired posterior distribution when the number of iterations tends to infinity. For this reason, we always perform a burn-in phase whose draws are discarded from the results. However, there is no systematic way of deciding when to stop the burn-in and start collecting MCMC samples. Not knowing the shape of the posterior distribution in advance, we are unable to formally assess whether we are already sampling from it. Only rough diagnostics are available: the **coda** package (Plummer et al., 2006) provides several, both numeric and graphical. In some cases, several weeks of computations are needed to obtain reliable results!

- MCMC algorithms are iterative, so they need some initial values from which to start: this is called the *initialisation* of the algorithm. If they are not provided by the user, Stan will generate them automatically. In the case of complicated models with a large number of nodes, choosing how to initialise an MCMC sampler is far from trivial; a simple solution is to use their empirical estimates.

- MCMC algorithms are pseudo-random algorithms in the sense that they are based on the generation of pseudo-random numbers. Hence it is crucial to save the seed which initialises the random number generator to be able to reproduce and investigate their output.

7.3 Other Software Packages

7.3.1 BayesiaLab

BayesiaLab is produced by Bayesia (Laval, France) and provides an integrated workspace to handle BNs. Its main feature is a user-friendly and powerful graphical interface, which is essential to make BNs accessible to people without programming skills. Numerous types of plots and convenient reports can be generated. It can also import BN objects from other software and export Markov blankets to external programming languages.

The main goals of BayesiaLab are the translation of expert knowledge into a BN, the discovery of structure from data sets, and combining both to make the best possible use of available information. Temporal aspects can also be incorporated in the BN, and causal modelling for decision making can be performed by including specialised nodes. Even if some standard continuous distributions can be used to define the categories of the nodes, BNs are basically restricted to discrete variables.

In addition, BayesiaLab implements several algorithms to optimise the layout of the nodes for better visualising DAGs, and supports interactive manipulation of the proposed layouts.

Written in Java, it can be installed on all main platforms. A limited 30-day trial version can be downloaded from

7.3.2   Hugin

Hugin is a software package commercialised by Hugin Expert A/S
and developed in collaboration with researchers from Aalborg University
(Denmark). Its first release was a command-line tool created in the context
of an ESPRIT project (Andersen et al., 1989) whose final outcome was the
MUNIN expert system (Andreassen et al., 1989).

Modern versions of Hugin provide a graphical interface that allows
users to perform BN learning and inference without the need of learning a
programming language. It supports decision trees, discrete BNs, GBNs and
hybrid BNs assuming a conditional Gaussian distribution. Exact inference
is implemented using junction trees (Algorithm 6.4), and is complemented
by a sensitivity analysis of the resulting posterior probabilities. Furthermore,
missing values are supported both in learning and inference using the EM
algorithm.

A demo version of Hugin can be freely downloaded from

https://www.hugin.com/index.php/hugin-lite,

but it is limited to handle at most 50 states and learn BNs from samples of
at most 500 cases. An interface to R is provided by package **RHugin**, which
can be downloaded from:

http://rhugin.r-forge.r-project.org

7.3.3   GeNIe

GeNIe has been developed by Marek Druzdzel's Decision Systems Laboratory
at University of Pittsburgh, and was first released in the mid-1990s. It has
an intuitive and interactive graphical user interface that does not require any
programming skills. Its main features are:

- It supports discrete BNs, GBNs, CGBNs and hybrid BNs with no limitations
  on the functional form of the local distribution like those we covered in
  Chapter 5, as well as influence diagrams and DBNs with higher-order
  temporal influences.

- It provides two state-of-the-art sampling algorithms for approximate
  inference: EPIS-BN (Evidence Pre-propagation Importance Sampling; Yuan
  and Druzdzel, 2003) and AIS-BN (Adaptive Importance Sampling; Cheng
  and Druzdzel, 2000).

- It also offers a selection of structure learning algorithms including the PC algorithm and two greedy search-based algorithms; BN classifiers like naïve Bayes, Tree-Augmented naïve Bayes (TAN) and the Adaptive Bayes Network (ABN) models.

- It can learn the parameters of a BN or a DBN from incomplete data using the Expectation-Maximisation (EM) algorithm.

GeNIe was originally written for Windows but it now runs smoothly on macOS and Linux through Wine. The entire functionality of GeNIe can be accessed from several programming languages (C++ and C, .NET, .COM, Java, Python and R) through the SMILE (Structural Modeling, Inference and Learning Engine) cross-platform C++ library. SMILE serves as the backend of several other applications that complement GeNIe, such as BayesBox (an interactive online BN repository), BayesMobile (an iOS app for diagnostic applications) and QGeNIe (a rapid prototyping tool for building BNs from qualitative expert knowledge). Future development plans include support for geographic information systems (GIS).

While GeNIe and SMILE have recently become commercial products available from BayesFusion, LLC (https://www.bayesfusion.com), they are still available free of charge for academic research and teaching use. Fully functional 30-day evaluation versions of the programs are available for download for other uses.

8

Real-World Applications of Bayesian Networks

We will now apply the concepts we introduced in the previous chapters to the analysis of two real-world data sets from the life sciences: a protein-signalling data set from which we want to discover the interactions and the pathways characterising some biological processes in human cells, and a medical diagnostic data set which we will use as a basis to predict a human's body composition.

8.1 Learning Protein-Signalling Networks

BNs provide a versatile tool for the analysis of many kinds of biological data such as *single-nucleotide polymorphism* (SNP) data and *gene expression* profiles. Following the work of Friedman et al. (2000), the expression level or the allele frequency of each gene is associated with one node. In addition, we can include additional nodes denoting other attributes that affect the system, such as experimental conditions, temporal indicators and exogenous cellular conditions. As a result, we can model in a single, comprehensive BN both the biological mechanisms we are interested in and the external conditions influencing them at the same time. Some interesting applications along these lines are studied, among others, in Yu et al. (2004); Zou and Conzen (2005); Morota et al. (2012); Villa-Vialaneix et al. (2013); Hartley and Sebastiani (2013).

An outstanding example of how such data can be analysed effectively is presented in Sachs et al. (2005) using protein-signalling data. BNs were used to represent complex direct and indirect relationships among multiple interacting molecules while accommodating biological noise. The data consist in the simultaneous measurements of 11 phosphorylated proteins and phospholipids derived from thousands of individual primary immune system cells, subjected to both general and specific molecular interventions. The former ensure that the relevant signalling pathways are active, while the latter make causal inference possible by elucidating arc directions through stimulatory cues and inhibitory interventions.

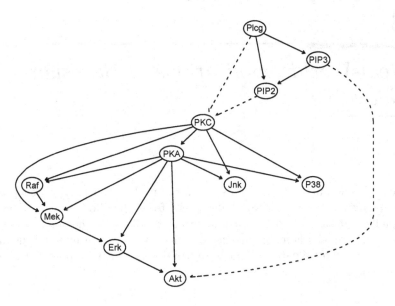

Figure 8.1
Protein-signalling network from Sachs et al. (2005). Signalling pathways that
are known from literature but were not captured by the BN are shown with
a dashed line.

The analysis described in Sachs et al. (2005) can be summarised as follows.

1. Outliers were removed and the data were discretised using the
   approach described in Hartemink (2001) because the distributional
   assumptions required by GBNs did not hold.

2. Structure learning was performed multiple times. In this way, a larger
   number of DAGs were explored in an effort to reduce the impact of
   locally optimal (but globally suboptimal) networks on learning and
   subsequent inference.

3. The DAGs learned in the previous step were averaged to produce
   a more robust model. This practice, known as *model averaging*, is
   known to result in a better predictive performance than choosing a
   single, high-scoring BN. The averaged DAG was created using the
   arcs present in at least 85% of the DAGs. This proportion measures
   the *strength* of each arc and provides the means to establish its
   significance given a *threshold* (85% in this case).

4. The validity of the averaged BN was evaluated against well-
   established signalling pathways from the literature.

The final BN is shown in Figure 8.1. It includes 11 nodes, one for each protein and phospholipid: PKC, PKA, Raf, Mek, Erk, Akt, Jnk, P38, Plcg (standing for Plcγ), PIP2 and PIP3. Every arc in that BN has been successfully validated, with very few exceptions:

- the arc between Pclg and PIP3 should be oriented in the other direction;

- there is a missing arc from PIP3 to Akt;

- there are two missing arcs from Plcg and PIP2 to PKC.

We will reproduce the whole analysis in the remainder of this section, using **bnlearn** and other packages covered in Section 7.1 to integrate missing functionality.

8.1.1 A Gaussian Bayesian Network

For the moment, we will consider only the 853 data manipulated with general interventions (*i.e.*, the observational data); we will investigate the complete data set (*i.e.*, both the observational and the interventional data) in Section 8.1.5.

```
library(bnlearn)
sachs <- read.table("sachs.data.txt", header = TRUE)
head(sachs)
    Raf   Mek  Plcg  PIP2  PIP3   Erk  Akt PKA   PKC  P38  Jnk
1 26.4 13.20  8.82 18.30 58.80  6.61 17.0 414 17.00 44.9 40.0
2 35.9 16.50 12.30 16.80  8.13 18.60 32.5 352  3.37 16.5 61.5
3 59.4 44.10 14.60 10.20 13.00 14.90 32.5 403 11.40 31.9 19.5
4 73.0 82.80 23.10 13.50  1.29  5.83 11.8 528 13.70 28.6 23.1
5 33.7 19.80  5.19  9.73 24.80 21.10 46.1 305  4.66 25.7 81.3
6 18.8  3.75 17.60 22.10 10.90 11.90 25.7 610 13.70 49.1 57.8
```

The data are continuous, as they represent the concentration of the molecules under investigation. Therefore, the standard approach in the literature is to assume the concentrations follow a Gaussian distribution and to use a GBN to build the protein-signalling network.

However, the DAGs learned under this parametric assumption are not satisfactory. For example, using Inter-IAMB we obtain a DAG with only 8 arcs (compared to the 17 in Figure 8.1) and only 2 of them are directed.

```
dag.iamb <- inter.iamb(sachs, test = "cor")
narcs(dag.iamb)
[1] 8
directed.arcs(dag.iamb)
     from  to
[1,] "P38" "PKC"
[2,] "Jnk" "PKC"
```

Other combinations of constraint-based algorithms and conditional independence tests, such as gs with test = "mc-cor", return the same DAG. The same is true for score-based and hybrid algorithms. If we compare dag.iamb with the DAG from Figure 8.1 (minus the arcs that were missed in the original paper), we can see that they have completely different structures.

```
sachs.modelstring <-
  paste0("[PKC][PKA|PKC][Raf|PKC:PKA][Mek|PKC:PKA:Raf][Erk|Mek:PKA]",
         "[Akt|Erk:PKA][P38|PKC:PKA][Jnk|PKC:PKA][Plcg][PIP3|Plcg]",
         "[PIP2|Plcg:PIP3]")
dag.sachs <- model2network(sachs.modelstring)
unlist(compare(dag.sachs, dag.iamb))
 tp fp fn
  0  8 17
```

Comparing the two DAGs again, but disregarding arc directions, reveals that some of the dependencies are correctly identified by inter.iamb but their directions are not.

```
unlist(compare(skeleton(dag.sachs), skeleton(dag.iamb)))
 tp fp fn
  8  0  9
```

The reason for the discrepancy between dag.sachs and dag.iamb is apparent from a graphical exploratory analysis of the data. Firstly, the empirical distributions of the molecules' concentrations are markedly different from a normal distribution. As an example, we plotted the distributions of Mek, P38, PIP2 and PIP3 in Figure 8.2. They are all strongly skewed because concentrations are positive numbers but a lot of them are small, and therefore cluster around zero. As a result, the variables are not symmetric and clearly violate the distributional assumptions underlying GBNs. Secondly, the dependence relationships in the data are not always linear; this is the case, as shown in Figure 8.3, for PKA and PKC. Most conditional independence tests and network scores designed to capture linear relationships have very low power in detecting nonlinear ones. In turn, structure learning algorithms using such statistics are unable to correctly learn the arcs in the DAG.

8.1.2   Discretising Gene Expressions

Since we have concluded that GBNs are not appropriate for the Sachs et al. (2005) data, we must now consider some alternative parametric assumptions. One possibility could be to explore monotone transformations like the logarithm. Another possibility would be to specify an appropriate conditional distribution for each variable, thus obtaining a hybrid network like those we explored in Chapter 5. However, this approach requires substantial prior knowledge on the signalling pathways, which may or may not be available. In the case of Sachs et al. (2005), such information was indeed available from

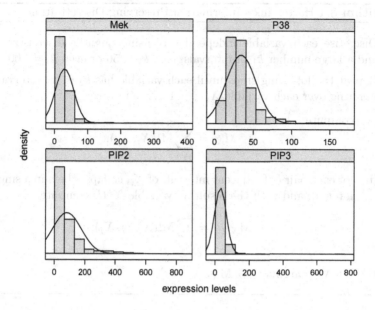

Figure 8.2
Densities of Mek, P38, PIP2 and PIP3. Histograms represent the empirical densities computed from the sachs data, while the curves are normal density functions with the appropriate means and variances. Note that on the right side of each histogram there are almost imperceptible bars, justifying the range of the x-axis and making the normal distribution irrelevant.

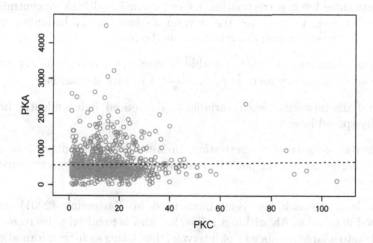

Figure 8.3
The (nonlinear) relationship between PKA and PKC; the dashed line is the fitted regression line for PKA against PKC.

Algorithm 8.1 Hartemink's Information-Preserving Discretisation

---

1. Discretise each variable independently using quantile discretisation and a large number k_1 of intervals, e.g., $k_1 = 50$ or even $k_1 = 100$.

2. Repeat the following steps until each variable has $k_2 \ll k_1$ intervals, iterating over each variable X_i, $i = 1, \ldots, p$ in turn:

   (a) compute
   $$\mathrm{M}_{X_i} = \sum_{j \neq i} \mathrm{MI}(X_i, X_j);$$

   (b) for each pair l of adjacent intervals of X_i, collapse them in a single interval, and with the resulting variable $X_i^*(l)$ compute
   $$\mathrm{M}_{X_i^*(l)} = \sum_{j \neq i} \mathrm{MI}(X_i^*(l), X_j);$$

   (c) set $X_i = \mathrm{argmax}_{X_i(l)} \mathrm{M}_{X_i^*(l)}$.

---

literature. However, the aim of the analysis was to use BNs as an automated probabilistic method to verify such information, not to build a BN with prior information and use it as an expert system.

Therefore, Sachs et al. (2005) decided to *discretise* the data and to model them with a discrete BN, which can accommodate skewness and nonlinear relationships at the cost of losing the ordering information. Since the variables in the BN represent concentration levels, it is intuitively appealing to discretise them into three levels corresponding to *low*, *average* and *high* concentrations.

To that end, we can use the `discretize` function in **bnlearn**, which implements three common discretisation methods:

- *quantile discretisation*: each variable is discretised independently into k intervals delimited by its $0, \frac{1}{k}, \frac{2}{k}, \ldots, \frac{(k-1)}{k}, 1$ empirical quantiles;

- *interval discretisation*: each variable is discretised independently into k equally-spaced intervals;

- *information-preserving discretisation*: variables are jointly discretised while preserving as much of the pairwise mutual information between the variables as possible.

The last approach has been introduced by Hartemink (2001) and is described in detail in Algorithm 8.1. The key idea is to initially discretise each variable into a large number k_1 of intervals, thus losing as little information as possible. Subsequently, the algorithm iterates over the variables and collapses, for each of them, the pair of adjacent intervals that minimises the loss of pairwise mutual information. The algorithm stops when all variables have

$k_2 \ll k_1$ intervals left. The resulting set of discretised variables reflects the dependence structure of the original data much better than either quantile or interval discretisation would allow because the discretisation takes pairwise dependencies into account. Clearly some information is always lost in the process; for instance, higher-level dependencies are completely disregarded and therefore are not likely to be preserved.

Algorithm 8.1 is implemented in the discretize function (method = "hartemink") along with quantile (method = "quantile") and interval discretisation (method = "interval").

```
dsachs <- discretize(sachs, method = "hartemink", breaks = 3,
            ibreaks = 60, idisc = "quantile")
```

The relevant arguments are idisc and ibreaks, which control how the data are initially discretised, and breaks, which specifies the number of levels of each discretised variable. ibreaks corresponds to k_1 in Algorithm 8.1, while breaks corresponds to k_2. Choosing good values for these arguments is a trade-off between quality and speed: high values of ibreaks preserve the characteristics of the original data to a greater extent, whereas smaller values result in much smaller memory usage and shorter running times.

8.1.3 Model Averaging

Following the analysis in Sachs et al. (2005), we can improve the quality of the structure learned from the data by averaging multiple CPDAGs. One possible approach to that end is to apply bootstrap resampling to dsachs and learn a set of 500 network structures.

```
boot <- boot.strength(dsachs, R = 500, algorithm = "hc",
            algorithm.args = list(score = "bde", iss = 10))
```

In the code above we learn a CPDAG with hill-climbing from each of the R bootstrap samples. Each variable in dsachs is now a factor with three levels, so we use the BDe score with a very low imaginary sample size.

The boot object returned by boot.strength is a data frame containing the strength of all possible arcs (in the strength column) and the probability of their direction (in the direction column) conditional on the fact that the from and to nodes are connected by an arc. This structure makes it easy to select the most significant arcs, as below.

```
boot[(boot$strength > 0.85) & (boot$direction >= 0.5), ]
     from   to strength direction
1     Raf  Mek    1.000  0.513000
23   Plcg PIP2    1.000  0.515000
24   Plcg PIP3    1.000  0.525000
34   PIP2 PIP3    1.000  0.511000
56    Erk  Akt    1.000  0.554000
57    Erk  PKA    0.990  0.560606
```

```
 67    Akt   PKA    1.000   0.560000
 89    PKC   P38    1.000   0.509000
 90    PKC   Jnk    1.000   0.507000
100    P38   Jnk    0.946   0.507400
```

Arcs are considered significant if they appear in at least 85% of the networks, and in the direction that appears most frequently. Arcs that are score equivalent in all the CPDAGs are considered to appear 50% of the time in each direction. Since all the values in the direction column above are close to 0.5, we can infer that the direction of the arcs is not well established and that they are probably all score equivalent. Interestingly, lowering the threshold from 85% to 50% does not change the results of the analysis, which seems to indicate that in this case the results are not sensitive to its value.

Having computed the significance for all possible arcs, we can now build the averaged network using averaged.network and the 85% threshold.

```
avg.boot <- averaged.network(boot, threshold = 0.85)
```

The averaged network avg.boot contains the same arcs as the network learned from the observational data in Sachs et al. (2005), which is shown in Figure 8.4. Even though the probability of the presence of each arc and of its possible directions are computed separately in boot.strength, we are not able to determine with any confidence which direction has better support from the discretised data. Therefore, we remove the directions from the arcs in avg.boot, which amounts to constructing its skeleton.

```
avg.boot <- skeleton(avg.boot)
```

As an alternative, we can average the results of several hill-climbing searches, each starting from a different DAG. Such DAGs can be generated randomly from a uniform distribution over the space of connected graphs with the MCMC algorithm proposed by Ide and Cozman (2002) and implemented in random.graph as method = "ic-dag". This ensures that no systematic bias is introduced in the learned DAGs. In addition, keeping only one randomly generated DAG every 50 ensures that the DAGs are different from each other so that the search space is covered as thoroughly as possible.

```
nodes <- names(dsachs)
start <- random.graph(nodes, method = "ic-dag", num = 500, every = 50)
netlist <- lapply(start, function(net) {
  hc(dsachs, score = "bde", iss = 10, start = net)
})
```

After using lapply to iterate over the DAGs in the start list and to pass each of them to hc with the start argument, we obtain a list of bn objects, which we call netlist. We can pass such a list to the custom.strength function to obtain a data frame with the same structure as those returned by boot.strength.

```
rnd <- custom.strength(netlist, nodes = nodes)
```

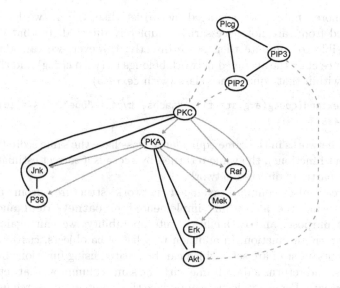

Figure 8.4
The undirected graph learned from the observational data in Sachs et al.
(2005). Missing arcs from the DAG in Figure 8.1 are plotted in grey.

```
rnd[(rnd$strength > 0.85) & (rnd$direction >= 0.5), ]
      from   to strength direction
1      Raf  Mek        1     0.510
33    PIP2 Plcg        1     0.503
34    PIP2 PIP3        1     0.501
43    PIP3 Plcg        1     0.502
57     Erk  PKA        1     0.503
66     Akt  Erk        1     0.504
67     Akt  PKA        1     0.507
99     P38  PKC        1     0.522
100    P38  Jnk        1     0.508
109    Jnk  PKC        1     0.514
avg.start <- averaged.network(rnd, threshold = 0.85)
```

The arcs identified as significant with this approach are the same as in
avg.boot (even though some are reversed), thus confirming the stability of
the averaged network obtained from bootstrap resampling. Arc directions
are again very close to 0.5, to the point we can safely disregard them. A
comparison of the equivalence classes of avg.boot and avg.start shows that
the two networks are equivalent.

```
all.equal(cpdag(avg.boot), cpdag(avg.start))
[1] TRUE
```

Furthermore, note how averaged networks, like the networks they are computed from, are not necessarily completely directed. In that case, it is not possible to compute their score directly. However, we can identify the equivalence class the averaged network belongs to (with cpdag) and then select a DAG within that equivalence class (with cextend).

```
score(cextend(cpdag(avg.start)), dsachs, type = "bde", iss = 10)
[1] -8498.88
```

Since all networks in the same equivalence class have the same score (for score-equivalent functions), the value returned by score is a correct estimate for the original, partially directed network.

We can also compute averaged network structures from bootstrap samples using the algorithms implemented in **catnet**, **deal** and **pcalg**. For this purpose, and to facilitate interoperability, we can again use the custom.strength function. In addition to a list of bn objects, custom.strength also accepts a list of arc sets, which can be created using functions from other packages, and returns a data frame with the same columns as that returned by boot.strength. For example, we can replace the hill-climbing search used above with the simulated annealing search implemented in **catnet** as follows.

```
library(catnet)
netlist <- vector(500, mode = "list")
ndata <- nrow(dsachs)
nodes <- names(dsachs)
netlist <- lapply(netlist, function(net) {
  boot <- dsachs[sample(ndata, replace = TRUE), ]
  top.ord <- cnSearchOrder(boot)
  best <- cnFindBIC(top.ord, ndata)
  cnMatEdges(best)
})
sann <- custom.strength(netlist, nodes = nodes)
```

The code above is similar to that used to create avg.start. After creating an empty list with 500 slots to hold the arc sets, we iterate over it using lapply. At each iteration:

1. we create the bootstrap sample boot by subsetting the original data frame dsachs with the sample function and replace = TRUE;

2. we learn the topological ordering top.ord of the nodes from the bootstrap sample using cnSearchOrder from **catnet**;

3. we learn (again from the data) the DAG with the best BIC score given top.ord using cnFindBIC from **catnet**;

4. we extract the arcs from best using cnMatEdges from **catnet**.

Finally, we perform model averaging with custom.strength from **bnlearn**. The result is stored in the sann object, which we can investigate as before.

```
sann[(sann$strength > 0.85) & (sann$direction >= 0.5), ]
     from   to strength direction
1     Raf  Mek    1.000        0.5
11    Mek  Raf    1.000        0.5
23   Plcg PIP2    0.986        0.5
33   PIP2 Plcg    0.986        0.5
34   PIP2 PIP3    1.000        0.5
44   PIP3 PIP2    1.000        0.5
56    Erk  Akt    1.000        0.5
66    Akt  Erk    1.000        0.5
67    Akt  PKA    1.000        0.5
77    PKA  Akt    1.000        0.5
89    PKC  P38    1.000        0.5
90    PKC  Jnk    1.000        0.5
99    P38  PKC    1.000        0.5
109   Jnk  PKC    1.000        0.5
avg.catnet <- averaged.network(sann, threshold = 0.85)
```

Unlike avg.boot and avg.start, avg.catnet is clearly an undirected graph; both directions have a probability of exactly 0.5 for every arc because the 500 DAGs learned from the bootstrap samples do not contain any v-structure. Furthermore, we can see with the narcs function that avg.catnet only has 7 arcs compared to the 10 of avg.start.

```
narcs(avg.catnet)
[1] 7
narcs(avg.start)
[1] 10
```

avg.catnet also seems not to fit the data very well: the BDe score returned by score for the same iss we used in structure learning is higher for avg.start than for avg.catnet.

```
score(cextend(cpdag(avg.catnet)), dsachs, type = "bde", iss = 10)
[1] -8548.34
score(cextend(cpdag(avg.start)), dsachs, type = "bde", iss = 10)
[1] -8498.88
```

Such differences can be attributed to the different scores and structure learning algorithms used to build the sets of high-scoring networks. In particular, it is very common for arc directions to differ between learning algorithms.

8.1.4 Choosing the Significance Threshold

The value of the threshold above which an arc is considered significant, which is called the *significance threshold*, does not seem to have a huge influence on the analysis in Sachs et al. (2005). In fact, any value between 0.5 and 0.85 yields exactly the same results. So, for instance:

```
all.equal(averaged.network(boot, threshold = 0.50),
          averaged.network(boot, threshold = 0.70))
```
```
 [1] TRUE
```

The same holds for avg.catnet and avg.start. However, this is often not the case. Therefore, it is important to use a statistically motivated algorithm for choosing a suitable threshold instead of relying on ad-hoc values.

A solution to this problem is presented in Scutari and Nagarajan (2013), and implemented in **bnlearn** as the default value for the threshold argument in averaged.network. It is used when we do not specify threshold ourselves as in the code below.

The value of the threshold is computed as follows. If we denote the arc strengths stored in boot as $\hat{\mathbf{p}} = \{\hat{p}_i, i = 1, \ldots, k\}$, and $\hat{\mathbf{p}}_{(\cdot)}$ is

$$\hat{\mathbf{p}}_{(\cdot)} = \{0 \leqslant \hat{p}_{(1)} \leqslant \hat{p}_{(2)} \leqslant \ldots \leqslant \hat{p}_{(k)} \leqslant 1\}, \tag{8.1}$$

then we can define the corresponding arc strength in the (unknown) averaged network $G = (\mathbf{V}, A_0)$ as

$$\tilde{p}_{(i)} = \begin{cases} 1 & \text{if } a_{(i)} \in A_0 \\ 0 & \text{otherwise} \end{cases}, \tag{8.2}$$

that is, the set of strengths that characterises any arc as either significant or non-significant without any uncertainty. In other words,

$$\tilde{\mathbf{p}}_{(\cdot)} = \{0, \ldots, 0, 1, \ldots, 1\}. \tag{8.3}$$

The proportion t of elements of $\tilde{\mathbf{p}}_{(\cdot)}$ that are equal to 1 determines the number of arcs in the averaged network, and is a function of the significance threshold we want to estimate. One way to do that is to find the value $\hat{t}$ that minimises the L_1 norm

$$L_1\left(t; \hat{\mathbf{p}}_{(\cdot)}\right) = \int \left| F_{\hat{\mathbf{p}}_{(\cdot)}}(x) - F_{\tilde{\mathbf{p}}_{(\cdot)}}(x; t) \right| \, dx \tag{8.4}$$

between the cumulative distribution functions of $\tilde{\mathbf{p}}_{(\cdot)}$ and $\hat{\mathbf{p}}_{(\cdot)}$, and then to include every arc that satisfies

$$a_{(i)} \in A_0 \iff \hat{p}_{(i)} > F_{\hat{\mathbf{p}}_{(\cdot)}}^{-1}(\hat{t}) \tag{8.5}$$

in the average network. This amounts to finding the averaged network whose arc set is "closest" to the arc strength computed from the data, with $F_{\hat{\mathbf{p}}_{(\cdot)}}^{-1}(\hat{t})$ acting as the significance threshold.

For the dsachs data, the estimated value for the threshold is 0.358: any arc with a strength value strictly greater than that is considered significant. The resulting averaged network is similar to that obtained with the 85% threshold from Sachs et al. (2005), and it belongs to the same equivalence class as avg.boot.

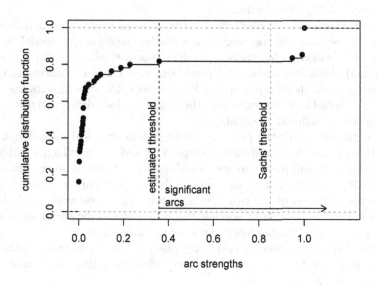

Figure 8.5
Cumulative distribution function of the arc strength values computed with bootstrap resampling from dsachs. The vertical dashed lines correspond to the estimated (black) and ad-hoc (grey) significance thresholds.

```
unlist(compare(cpdag(dag.sachs), cpdag(averaged.network(boot))))
 tp fp fn
  9  1  8
all.equal(cpdag(avg.boot), cpdag(averaged.network(boot)))
 [1] TRUE
```

The reason for the insensitivity of the averaged network to the value of the threshold is apparent from the plot of $F_{\hat{p}_{(\cdot)}}$ in Figure 8.5: arcs that are well supported by the data are clearly separated from those that are not, with the exception of one arc that falls right on the threshold. The arc with smallest strength above the threshold has strength 0.946, so any threshold that falls between 0.946 and 0.358 results in the same averaged network.

8.1.5 Handling Interventional Data

Usually, all the observations in a sample are collected under the same general conditions. This is true both for *observational data*, in which treatment allocation is outside the control of the investigator, and for *experimental data*, which are collected from randomised controlled trials. As a result, the sample can be easily modelled with a single BN, because all the observations follow the same probability distribution.

However, this is not the case when several samples resulting from different experiments are analysed together with a single, encompassing model. Such an approach is called *meta analysis* (see Kulinskaya et al., 2008, for a gentle introduction). First, environmental conditions and other exogenous factors may differ between those experiments. Furthermore, the experiments may be different in themselves: for example, they may explore different treatment regimes or target different populations.

This is the case with the protein-signalling data analysed in Sachs et al. (2005). In addition to the data set we have analysed so far, which is subject only to general stimuli meant to activate the desired paths, 9 other data sets subject to different stimulatory cues and inhibitory interventions were used to elucidate the direction of the causal relationships in the network. Such data are often called *interventional* because the values of specific variables in the model are set by an external intervention of the investigator.

Overall, the 10 data sets contain 5, 400 observations; in addition to the 11 signalling levels analysed above, the protein which is activated or inhibited (INT) is recorded for each sample.

```
isachs <- read.table("sachs.interventional.txt", header = TRUE,
            colClasses = "factor")
```

One intuitive way to model these data sets with a single, encompassing BN is to include the intervention INT in the network and to make all variables depend on it. This can be achieved with a whitelist containing all possible arcs from INT to the other nodes, thus forcing these arcs to be present in the learned DAG.

```
wh <- matrix(c(rep("INT", 11), names(isachs)[1:11]), ncol = 2)
dag.wh <- tabu(isachs, whitelist = wh, score = "bde",
            iss = 10, tabu = 50)
```

Using tabu search instead of hill-climbing improves the stability of the score-based search: once a locally optimum DAG is found, tabu search performs an additional 50 iterations (as specified by the tabu argument) to ensure that no other (and potentially better) local optimum is found.

We can also let the structure learning algorithm decide which arcs connecting INT to the other nodes should be included in the DAG. To this end, we can use the tiers2blacklist function to blacklist all arcs toward INT, thus ensuring that only outgoing arcs can be included in the DAG. In the general case, tiers2blacklist builds a blacklist such that all arcs going from a node in a particular element of the nodes argument to a node in one of the previous elements are blacklisted.

```
tiers <- list("INT", names(isachs)[1:11])
bl <- tiers2blacklist(tiers)
```

```
dag.tiers <- tabu(isachs, blacklist = bl, score = "bde",
                  iss = 1, tabu = 50)
```

The BNs learned with these two approaches are shown in the top two panels of Figure 8.6. Some of the structural features detected in Sachs et al. (2005) are present in both dag.wh and dag.tiers. For example, the interplay between Plcg, PIP2 and PIP3 and between PKC, P38 and Jnk are both modelled correctly. The lack of any direct intervention on PIP2 is also correctly modelled in dag.tiers. The most noticeable feature missing from both DAGs is the pathway linking Raf to Akt through Mek and Erk.

The approach used in Sachs et al. (2005) yields much better results. Instead of including the interventions in the network as an additional node, they used a modified BDe score (labelled "mbde" in **bnlearn**) incorporating the effects of the interventions into the score components associated with each node (Cooper and Yoo, 1999). When we are controlling the value of a node experimentally, its value is not determined by the other nodes in the BN, but by the experimenter's intervention. Accordingly, mbde disregards the effects of the parents on a controlled node for those observations that are subject to interventions (on that node) while otherwise behaving as the standard bde for other observations.

Since the value of INT identifies which node is subject to either a stimulatory cue or an inhibitory intervention for each observation, we can easily construct a named list of which observations are manipulated for each node.

```
INT <- sapply(1:11, function(x) which(isachs$INT == x))
nodes <- names(isachs)[1:11]
names(INT) <- nodes
```

We can then pass this list to tabu as an additional argument for mbde. In addition, we combine the use of mbde with model averaging and random starting points as discussed in Section 8.1.3. To improve the stability of the averaged network, we generate the set of the starting networks for tabu using the algorithm from Melançon et al. (2001), which is not limited to connected networks as that from Ide and Cozman (2002). In addition, we actually use only 1 generated network every 100 to obtain a more diverse set.

```
start <- random.graph(nodes = nodes, method = "melancon", num = 500,
            burn.in = 10^5, every = 100)
netlist <- lapply(start, function(net) {
  tabu(isachs[, 1:11], score = "mbde", exp = INT, iss = 1,
    start = net, tabu = 50)
})
intscore <- custom.strength(netlist, nodes = nodes, cpdag = FALSE)
```

Note that we have set cpdag = FALSE in custom.strength, so that the DAGs being averaged are not transformed into CPDAGs before computing arc strengths and direction probabilities. The reason is that, unlike the BDe score

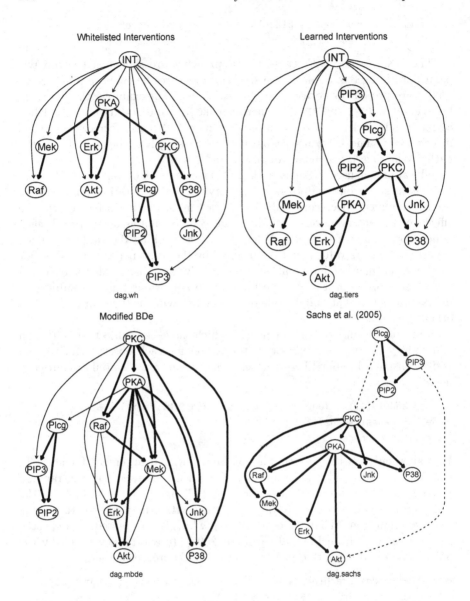

Figure 8.6

DAGs learned from isachs. The first two networks (dag.wh on the top left, dag.tiers on the top right) have been learned by including INT and adding arcs to model stimulatory cues and inhibitory interventions. The third BN (dag.mbde, on the bottom left) has been learned with model averaging and the mbde score; arcs plotted with a thicker line width make up the validated BN (bottom right) from Sachs et al. (2005).

we used in Section 8.1.3, mbde is not score equivalent as a consequence of incorporating the interventions. In fact, information about the interventions is a form of prior information, and makes it possible to identify arc directions even when the arc would be score equivalent from the data alone.

Averaging the DAGs with averaged.network and the threshold from Section 8.1.4, we are finally able to correctly identify all the arcs in the network from Sachs et al. (2005).

```
dag.mbde <- averaged.network(intscore)
unlist(compare(dag.sachs, dag.mbde))
 tp fp fn
 17  8  0
```

As we can see from Figure 8.6, dag.mbde is much closer to the validated network from Sachs et al. (2005) than any of the other BNs learned in this section. All the arcs from the validated network are correctly learned; in the output above we have 17 true positives and 0 false negatives, and the original network contains 17 arcs. The three arcs that were missing in Sachs et al. (2005) are missing in dag.mbde as well. The arcs from dag.mbde that are not present in the validated network were identified in Sachs et al. (2005) and discarded due to their comparatively low strength; this may imply that the simulated annealing algorithm used in Sachs et al. (2005) performs better on this data set than tabu search.

8.1.6   Querying the Network

In their paper, Sachs et al. (2005) used the validated BN to substantiate two claims:

1. a direct perturbation of Erk should influence Akt;
2. a direct perturbation of Erk should not influence PKA.

The probability distributions of Erk, Akt and PKA were then compared with the results of two ad-hoc experiments to confirm the validity and the direction of the inferred causal influences.

Given the size of the BN, we can perform the queries corresponding to the two claims above using any of the exact and approximate inference algorithms introduced in Section 6.6.2. First, we need to create a bn.fit object for the validated network structure from Sachs et al. (2005).

```
isachs <- isachs[, 1:11]
for (i in names(isachs))
  levels(isachs[, i]) = c("LOW", "AVG", "HIGH")
fitted <- bn.fit(dag.sachs, isachs, method = "bayes")
```

The INT variable, which codifies the intervention applied to each observation, is not needed for inference and is therefore dropped from the data set.

Furthermore, we rename the expression levels of each protein to make both the subsequent R code and its output more readable.

Subsequently, we can perform the two queries using the junction tree algorithm provided by the **gRain** package, which we have previously explored in Section 1.6.2.1. The results are shown in Figure 8.7.

```
library(gRain)
jtree <- compile(as.grain(fitted))
```

We can introduce the direct perturbation of Erk required by both queries by calling setEvidence as follows. In causal terms, this would be an ideal intervention.

```
jlow <- setEvidence(jtree, nodes = "Erk", states  = "LOW")
```

As we can see from the code below, the marginal distribution of Akt changes depending on whether we take the evidence (intervention) into account or not.

```
querygrain(jtree, nodes = "Akt")$Akt
Akt
   LOW    AVG   HIGH
0.6094 0.3104 0.0802
querygrain(jlow, nodes = "Akt")$Akt
Akt
     LOW      AVG     HIGH
0.666516 0.333333 0.000151
```

The slight inhibition of Akt induced by the inhibition of Erk agrees with both the direction of the arc linking the two nodes and the additional experiments performed by Sachs et al. (2005). In causal terms, the fact that changes in Erk affect Akt supports the existence of a causal link from the former to the latter.

As far as PKA is concerned, both the validated network and the additional experimental evidence support the existence of a causal link from PKA to Erk. Therefore, interventions to Erk cannot affect PKA. In a causal setting, interventions on Erk would block all biological influences from other proteins, which amounts to removing all the parents of Erk from the DAG.

```
causal.sachs <- drop.arc(dag.sachs, "PKA", "Erk")
causal.sachs <- drop.arc(causal.sachs, "Mek", "Erk")
cfitted <- bn.fit(causal.sachs, isachs, method = "bayes")
cjtree <- compile(as.grain(cfitted))
cjlow <- setEvidence(cjtree, nodes = "Erk", states  = "LOW")
```

After building the junction tree for this new DAG, called causal.sachs, we can perform the same query on both cjtree and cjlow as we did above.

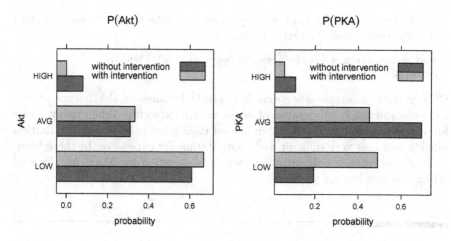

Figure 8.7
Probability distributions of Akt and PKA before and after inhibiting Erk, considered in a non-causal setting.

```
querygrain(cjtree, nodes = "PKA")$PKA
 PKA
   LOW   AVG  HIGH
 0.194 0.696 0.110
querygrain(cjlow, nodes = "PKA")$PKA
 PKA
   LOW   AVG  HIGH
 0.194 0.696 0.110
```

Indeed, PKA has exactly the same distribution in both cases. However, knowledge of the expression level of Erk may still alter our expectations on PKA if we treat it as evidence instead of an ideal intervention. In practice this implies the use of the original junction trees jtree and jlow, as opposed to the modified cjtree and cjlow we used for the previous query.

```
querygrain(jtree, nodes = "PKA")$PKA
 PKA
   LOW   AVG  HIGH
 0.194 0.696 0.110
querygrain(jlow, nodes = "PKA")$PKA
 PKA
    LOW    AVG   HIGH
 0.4897 0.4516 0.0586
```

All the queries illustrated above can be easily changed to maximum a posteriori queries by finding the largest element (with the which.max function) in the distribution of the target node.

```
names(which.max(querygrain(jlow, nodes = c("PKA"))$PKA))
[1] "LOW"
```

Clearly, such a simple approach is possible because of the nature of the evidence and the small number of nodes we are exploring. When many nodes are explored simultaneously, inference on their joint conditional distribution quickly becomes very difficult and computationally expensive. In these high-dimensional settings, algorithms specifically designed for MAP queries and ad-hoc approaches are preferable.

8.2   Predicting the Body Composition[1]

The *human body composition* is the distribution of the three components that form body weight: *bone*, *fat* and *lean*. They identify, respectively, the mineral content, the fat and the remaining mass of the body (mainly muscles). More in detail, body composition is measured separately across *trunk*, *legs* and *arms*, and it is an important diagnostic tool since the ratios of these masses can reveal regional physiological disorders. One of the most common ways to measure these masses is the dual-energy x-ray absorptiometry (DXA; Centers for Disease Control and Prevention, 2010), which is unfortunately time-consuming and very expensive. Therefore, there is an interest in alternative protocols that can be used for the same purpose.

In the following we will try, in a very simple way, to predict body composition from related quantities that are much cheaper and easier to measure: *age*, *height*, *weight* and *waist circumference*. To this end, we will use a sample of 100 white men collected by the NHANES project (Centers for Disease Control and Prevention, 2004) which includes simultaneous measurements for the variables above. This data set is available in the **rbmn** package under the name boco.

```
library(rbmn)
data(boco)
round(head(boco), 1)
   A   H     W   C   TF   LF  AF   TL   LL   AL  TB  LB  AB
1 83 182  92.6 117 17.1  8.9 3.0 31.2 18.5  6.6 0.6 1.1 0.5
2 68 169  74.7  93  8.3  5.4 2.0 28.0 16.2  7.5 0.7 1.0 0.5
3 28 182 112.2 112 17.7 11.3 3.1 36.7 24.5 10.1 0.8 1.1 0.5
4 41 171  82.6  96 10.6  6.5 1.8 29.2 19.6  7.8 0.8 1.1 0.5
5 85 169  71.1 102 10.9  4.7 1.9 26.2 14.5  5.8 0.6 1.0 0.4
6 29 176  88.4  96 11.2  7.5 2.7 31.4 19.8  8.3 0.7 0.9 0.4
```

---

[1]This section was written with the help of Simiao Tian (MIAJ, INRA, Jouy-en-Josas).

label	definition	unit
TB	Trunk Bone	kg
TF	Trunk Fat	kg
TL	Trunk Lean	kg
LB	Leg Bone	kg
LF	Leg Fat	kg
LL	Leg Lean	kg
AB	Arm Bone	kg
AF	Arm Fat	kg
AL	Arm Lean	kg
A	Age	year
W	Weight	kg
H	Height	cm
C	Waist Circumference	cm
B	Body Mass Index	kg/m^2

Table 8.1
The nine variables we are interested in predicting along with the possible covariates, their labels and their units of measurement.

```
boco$B <- boco$W / boco$H^2 * 10^4
dim(boco)
 [1] 100  14
n <- nrow(boco)
vr <- c("TF", "LF", "AF", "TL", "LL", "AL", "TB", "LB", "AB")
co <- c("A", "H", "W", "C", "B")
```

First, we computed the body mass index (B) of each individual and added it to boco. It is a very popular score, normalising the weight by the height: a weight of 100kg has very different health implications for a person 160cm tall than for another person that is 185cm tall. Therefore, we now have a set of 5 covariates and 9 variables of interest for a sample size of $n = 100$. Their labels and definitions are given in Table 8.1.

8.2.1 Aim of the Study

A standard statistical approach used for prediction is the multivariate multiple linear regression. Denoting the (response) variables of interest by $\mathbf{Y}$ and the covariates (the explanatory variables) by $\mathbf{X}$, the model has the following form:

$$\mathbf{Y} = \mathbf{X\Theta} + \mathbf{E} \tag{8.6}$$

$$V(vec(\mathbf{E})) = \mathbf{I}_n \otimes \mathbf{\Sigma} \tag{8.7}$$

where:

- **Y** is an $n \times p$ matrix;

- **X** is an $n \times (q + 1)$ matrix (including a **1** column for the intercept as well as the covariates);

- **Θ** is a $(q + 1) \times p$ matrix for the regression coefficients;

- **E** is an $n \times p$ matrix for the error terms;

- **Σ** is a $p \times p$ covariance matrix for the error terms;

- n is the number of observations;

- p is the number of variables; and

- q is the number of covariates.

The number of parameters is $p(q + 1)$ for the regression coefficients and $p(p + 1)/2$ for the covariance matrix of the error terms. When p and q are large, n must be larger still to provide enough predictive power. Of course, it is well known that covariances are more difficult to estimate from the data than mean values. Using BNs can be an effective way to dramatically reduce the number of parameters in the model and thus the required sample size. Specifically, a GBN defined over p variables and q covariates, with n_c arcs from the covariates to the variables and n_v arcs between the variables, will have only $2p + n_c + n_v$ parameters and a formulation that is consistent with Equations (8.6) and (8.7).

Indeed, these equations can be interpreted as a saturated GBN model (with all possible arcs present in the DAG) in which the roles of variables and covariates are emphasised. Therefore, we can reasonably expect better predictions from a sparse GBN than from a standard multivariate multiple regression when the sample size is too small for the latter. We will show an example of how we can learn such a GBN with the aim of predicting the nine variables with the five covariates in Table 8.1. It could be argued that W and H make B redundant, but their relationship is not linear and therefore they convey different information in a linear model.

8.2.2 Designing the Predictive Approach

8.2.2.1 *Assessing the Quality of a Predictor*

First we will split the data into a training set (dtrain) and a validation set (dval). The training set will be used to learn or estimate the models; the validation set will be used to select those models that perform the prediction of new individuals well, thus reducing the risk of overfitting (Hastie et al.,

2009). Here we will use a randomly generated split[2] into sets of equal size, each comprising 50 individuals.

```
set.seed(42)
sub <- split(sample(n), c("train", "validation"))
dtrain <- boco[sub$train, ]
dval <- boco[sub$validation, ]
```

We should also assess the bias-variance trade-off of a prediction. The discrepancy between the observed value and the predicted mean will capture the bias; the standard deviation will be given by that of the predictor. For each variable, the *standard error of prediction* (SEP) is defined by the classic formula

$$\text{SEP} = \sqrt{\text{bias}^2 + \text{standard deviation}^2}. \tag{8.8}$$

Another important consideration when choosing a model for prediction is its interpretability. When their statistical properties are comparable, models that elucidate the phenomenon under investigation are usually preferable to empirical, black-box ones. In the context of BNs, models with not too many arcs, and arcs with a physiological interpretation for body composition will be more appealing than the saturated model in Equations (8.6) and (8.7). In this respect, we hope that complex relationships between the variables will disappear after incorporating the good covariates. In other words: complex dependencies in the joint distribution of the variables may simplify or even disappear when conditioning on a covariate.

8.2.2.2 The Saturated BN

In order to have a term of comparison for the more parsimonious models we will fit later, we first consider the multiple regression equivalent of a saturated BN (with all possible arcs between variables and/or covariates).

```
saturated <- lm(cbind(TF, LF, AF, TL, LL, AL, TB, LB, AB) ~
                A + H + W + C + B, data = dtrain)
resid.df <- anova(saturated)["Residuals", "Df"]
preds <- predict(saturated, newdata = dval)
bias <- abs(dval[, vr] - preds)
stdev <- outer(rep(1, nrow(dtrain)),
        sqrt(colSums(residuals(saturated)^2)/resid.df), "*")
sep <- sqrt(bias^2 + stdev^2)
summary <- cbind("|Bias|" = colMeans(bias),
                "Sd.Dev" = colMeans(stdev),
                "SEP" = colMeans(sep))
```

---

[2]The splits are in fact generated using pseudo-random numbers, hence it is important for reproducibility to use a known random seed that should be set at the beginning of the analysis with set.seed. Using the same random number generation itself is clearly important as well: the default one changed in R 3.6.0.

```
round(summary, 2)
     |Bias| Sd.Dev  SEP
 TF   1.16   1.88  2.32
 LF   1.18   1.67  2.14
 AF   0.31   0.52  0.64
 TL   1.09   1.86  2.29
 LL   0.85   1.32  1.68
 AL   0.51   0.75  0.95
 TB   0.08   0.10  0.13
 LB   0.10   0.13  0.17
 AB   0.04   0.05  0.07
round(colSums(summary), 2)
 |Bias| Sd.Dev    SEP
   5.32   8.29  10.38
```

The output above shows a cumulated SEP of 10.38 over the 9 variables, with a
bias of 5.32 and a standard deviation of 8.29.[3] These three values do not satisfy
Equation (8.8), because we are summing SEPs for all predicted individuals and
variables, and the definition of SEP is not additive. The saturated GBN has
36 arcs between the variables and 45 arcs from the covariates to the variables.

8.2.2.3 Convenient BNs

The training set dtrain can also be used to learn a GBN with one of the
algorithms available in **bnlearn**, as below.[4]

```
library(bnlearn)
dag <- hc(dtrain)
bnlearn:::fcat(modelstring(dag))
    [A][H][W|H][B|H:W][C|A:H:B][TF|C:B][AF|W:TF][LF|A:AF]
    [TL|H:W:C:TF:LF][TB|C:TL][LB|TB][AB|LB]
    [LL|A:W:TF:LF:AF:TL:AB][AL|LL:TB:AB]
```

This basic attempt at learning a GBN does not necessarily produce a DAG
that is convenient to use for prediction. We would prefer to have covariates
as parents of the response variables that we are trying to predict so that we
can immediately obtain their conditional probability distribution. By chance,
this is the case for dag. However, even small changes in the code above such as
reversing the order of the variables in dtrain result in learning arcs like TF → C,
LF → B and AB → H. As we have seen in Chapter 6, we can use conditioning
other than that implied by a topological order but prediction would require
performing inference for each new observation.

---

[3]We choose to average the criteria over the variables to get a set of simple summaries to
describe each model as a whole. This is appropriate because all variables are on the same
scale and because they contribute equally to body weight. In more complex scenarios it may
be better to use the Mahalanobis distance (Mahalanobis, 1936).

[4]fcat is an internal function in **bnlearn** that is used by print to wrap long model strings.
It is not exported and may change in the future.

Blacklists and whitelists provide a simple way to enforce such constraints on the network structure when learning the GBN. The strongest constraint is to force every covariate to be the parent of every response variable; that means to include 5×9 arcs in the DAG.

```
wl1 <- cbind(from = rep(co, each = 9), to = rep(vr, 5))
dag.wl <- hc(dtrain, whitelist = wl1)
bnlearn:::fcat(modelstring(dag.wl))
    [A][H][W|H][B|H:W][C|A:H:B][LF|A:H:W:C:B]
    [AF|A:H:W:C:LF:B][LL|A:H:W:C:AF:B][TF|A:H:W:C:LL:B]
    [TB|A:H:W:C:AF:LL:B][AL|A:H:W:C:TF:LF:LL:TB:B]
    [LB|A:H:W:C:TF:TB:B][AB|A:H:W:C:AL:LB:B]
    [TL|A:H:W:C:TF:LF:AF:LL:AL:AB:B]
```

As expected, the covariates are in the first positions in the topological ordering of the DAG. To relax the constraints on structure learning, we can switch from the whitelist wl1 to a blacklist that only prevents unwanted arcs. In other words, variables cannot be the parents of covariates but covariates are not necessarily the parents of all the variables.

```
bl1 <- wl1[, 2:1]
dag.bl <- hc(dtrain, blacklist = bl1)
bnlearn:::fcat(modelstring(dag.bl))
    [A][H][W|H][B|H:W][C|A:H:B][TF|C:B][AF|W:TF][LF|A:AF]
    [TL|H:W:C:TF:LF][TB|C:TL][LB|TB][AB|LB]
    [LL|A:W:TF:LF:AF:TL:AB][AL|LL:TB:AB]
all.equal(dag.wl, dag.bl)
  [1] "Different number of directed/undirected arcs"
```

The resulting DAG is similar for the covariates but different for the response variables. In fact, every combination of the two lists is possible, and any overlap is handled properly by every algorithm in **bnlearn**.

```
iwl <- 1:15
wl2 <- wl1[iwl, ]
bl2 <- bl1[-iwl, ]
dag.wlbl <- hc(dtrain, whitelist = wl2, blacklist = bl2)
bnlearn:::fcat(modelstring(dag.wlbl))
    [A][H][W|H][B|H:W][C|A:H:B][TF|A:H:W:C][LF|A:H:TF:B]
    [TL|A:H:W:C:TF:LF][TB|A:C:TL][LB|A:TB][AB|A:LB]
    [AL|A:H:W:C:TL:TB:AB][AF|A:H:TF:LF:AL:TB]
    [LL|A:H:W:TF:LF:AF:TL:AL:AB]
```

It is important to note that we are not interested in the distribution of the covariates since we will assume them to be known for the purposes of prediction. Clearly, interactions between covariates should be allowed to some extent for the GBN to be correctly specified and to limit the bias in the regression coefficients associated with the covariates in the local distributions of the response variables. However, such interactions do not directly affect predictions. We are just interested in an efficient and simple form for the conditional distribution $Y_1, Y_2, \ldots, Y_p \mid X_1, X_2, \ldots, X_q$.

8.2.3   Looking for Candidate BNs

With these considerations in mind, we can implement a search strategy for good predictive BNs. We will use the SEP score from Equation (8.8) and its two components (the bias and the standard deviation) as measures of the BN's performance on the validation set dval. As a first step, we estimate the parameters of the model from the training set dtrain using the dag.wl object we created in the previous section.

```
bn2 <- bn.fit(dag.wl, data = dtrain)
```

Then we obtain the conditional distribution for each individual i in the validation set with **rbmn**, and we use them to compute bias and standard deviation. We store them in two data frames, bias and stde.

```
library(rbmn)
mvnorm.dist <- gema2mn(nbn2gema(bnfit2nbn(bn2)))
bias <- stdev <- dval[, vr]
for (i in seq(nrow(dval))) {
  mvnorm.cond <- condi4joint(mvnorm.dist, par = vr, pour = co,
                    unlist(dval[i, co]))
  bias[i, vr] <- dval[i, vr] - mvnorm.cond$mu[vr]
  stdev[i, vr] <- sqrt(diag(mvnorm.cond$gamma)[vr])
}#FOR
sep <- sqrt(bias^2 + stdev^2)
```

We can obtain a global score by summing over all the observations in the validation data set, as we did for the saturated model.

```
gscores <- cbind("|Bias|" = colMeans(abs(bias)),
                 "Sd.Dev" = colMeans(stdev),
                 "SEP" = colMeans(sep))
round(gscores, 2)
   |Bias| Sd.Dev SEP
TF   1.16   1.90 2.33
LF   1.18   1.67 2.14
AF   0.31   0.53 0.64
TL   1.09   1.93 2.35
LL   0.85   1.33 1.69
AL   0.51   0.80 0.99
TB   0.08   0.10 0.13
LB   0.10   0.13 0.17
AB   0.04   0.05 0.07
```

DAG 2

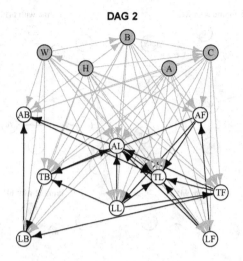

Figure 8.8
Network structure of dag.wl. Only the arcs we are interested in, those between two variables, are drawn in black; all the others are shown in grey. Nodes corresponding to covariates have a grey background.

```
round(colSums(gscores), 2)
|Bias| Sd.Dev    SEP
  5.32    8.44  10.51
```

These results are encouraging compared to those arising from the saturated GBN (5.32, 8.29, 10.38) if we consider that we are using fewer parameters (65 instead of 99). However, as previously mentioned, it is also important to check whether dag.wl is easy to interpret. For this purpose, we can plot dag.wl with a convenient layout for the nodes corresponding to the 3×3 variables and the 5 covariates. For brevity, we have prepared the matrix with the node coordinates in advance, and we load it below from the bc.poco.rda file.

```
library(igraph)
load("bc.poco.rda")
cbind(position, colour)[c(10:11, 1:3), ]
     x    y   colour
A  "70" "65" "lightgrey"
H  "30" "65" "lightgrey"
TF "95" "25" "white"
LF "90" "10" "white"
AF "85" "50" "white"
idag2 <- as.igraph(dag.wl)
nad <- V(idag2)$label <- V(idag2)$name
```

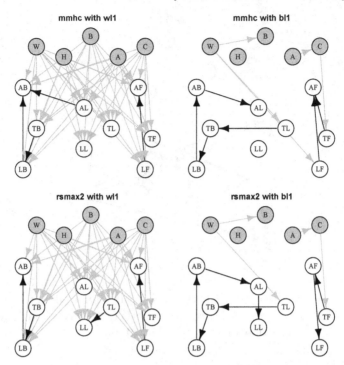

Figure 8.9
DAGs learned by mmhc (top left and top right) and rsmax2 (bottom left and
bottom right) with either the whitelist wl1 (top left and bottom left) or the
blacklist bl1 (top right and bottom right).

```
edge.col <- rep("lightgrey", narcs(dag.wl))
aa <- which((arcs(dag.wl)[, "from"] %in% vr) &
            (arcs(dag.wl)[, "to"] %in% vr))
va <- as.numeric(E(idag2, P = t(arcs(dag.wl)[aa, ])))
edge.col[va] <- "black"
plot(idag2, layout = position[nad, ], main = "DAG 2",
     edge.color = edge.col, vertex.color = colour[nad])
```

The plot produced by the code above is shown in Figure 8.8. No systematic
structure is apparent, and a sparser DAG with a comparable SEP score
would surely be preferable. Unfortunately, using the blacklist bl1 instead of
the whitelist wl1 does not improve things significantly. On the other hand,
changing structure learning from a score-based to a hybrid algorithm has a
dramatic effect, as can be seen in Figure 8.9.

    First of all, there are fewer arcs between the variables. In addition, the
two-way structure of the regions (trunk, legs, arms) × (lean, fat, bone) is
clearly recognisable. The GBN learned with rsmax2 and the whitelist looks

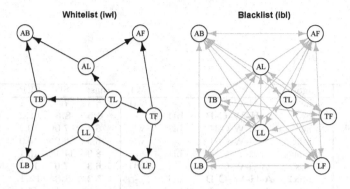

Figure 8.10
Arcs included in the whitelist (iwl, in black) and in the blacklist (ibl, in grey) used to look for a minimal subset of covariates.

quite promising since there are only a few arcs left between the variables, and they only link variables for the same type of mass. At the same time, all variables depend on all covariates due to the whitelist, which is not desirable when aiming for a parsimonious model.

Therefore, we wonder whether we need to use all the covariates to obtain good predictions. To investigate this further we introduce some restrictions on the arcs linking the variables, either with a whitelist (to see if more arcs are necessary) or with a blacklist (to prevent some arc from appearing); and we still blacklist arcs from a variable to a covariate. As the underlying theory, we assume that conditionally to the covariates:

- Lean and fat remain correlated (arcs between TL, LL, AL and TF, LF, AF).

- Lean and bone remain correlated (arcs between TL, LL, AL and TB, LB, AB).

- Fat and bone are conditionally independent from the lean mass (no arcs between TF, LF, AF and TB, LB, AB).

- Trunk weight influences leg weight since legs carry the body of which the trunk is the heaviest part (arcs between TB, TF, TL and LB, LF, LL).

- Trunk weight influences arm weight (arcs between TB, TF, TL and AB, AF, AL).

- There is no direct arc between LB, LF, LL and AB, AF, AL.

The arcs corresponding to these assumptions and included in the whitelist iwl or in the blacklist ibl are shown in Figure 8.10. The whitelist forces all the arcs with a clear interpretation to be present in the DAG, while allowing additional arcs. The blacklist prevents all the arcs that do not have a clear interpretation from being included in the DAG, regardless of their directions, and is designed to complement the whitelist.

	Algo.	Cova.	BL	WL	|Bias|	SD	SEP	c2v	v2v
dag.wl	hc	A-H-W-C-B	-	wl1	6.3	7.1	10.1	45	20
dag.bl	**hc**	**A-H-W-C-B**	**bl1**	**-**	**6.5**	**8.3**	**11.2**	**12**	**17**
dag.wlbl	hc	A-H-W-C-B	bl2	wl2	6.4	7.0	10.2	23	15
(1)	mmhc	A-H-W-C-B	-	wl1	6.3	6.9	10.0	45	3
(2)	mmhc	A-H-W-C-B	bl1	-	8.9	11.1	15.1	3	6
(3)	rsmax2	A-H-W-C-B	-	wl1	6.3	7.0	10.0	45	4
(4)	**rsmax2**	**A-H-W-C-B**	**bl1**	-	**7.2**	**8.8**	**12.1**	**3**	**6**
<1>	mmhc	W-H-B-A-C	ibl+rbl	-	8.9	11.1	15.1	3	3
<2>	**mmhc**	**W-H-B-A-C**	**rbl**	**iwl**	**7.3**	**9.1**	**12.4**	**2**	**15**
<3>	rsmax2	W-H-B-A-C	ibl+rbl	-	7.6	9.2	12.7	4	4
<4>	**rsmax2**	**W-H-B-A-C**	**rbl**	**iwl**	**7.3**	**9.1**	**12.4**	**2**	**15**
<5>	mmhc	W-H-A-C	ibl+rbl	-	9.6	12.4	16.7	3	3
<6>	**mmhc**	**W-H-A-C**	**rbl**	**iwl**	**7.2**	**9.0**	**12.2**	**3**	**15**
<7>	rsmax2	W-H-A-C	ibl+rbl	-	8.9	10.9	14.9	5	3
<8>	**rsmax2**	**W-H-A-C**	**rbl**	**iwl**	**7.2**	**9.0**	**12.2**	**3**	**15**
<9>	mmhc	W-H-C	ibl+rbl	-	9.6	12.4	16.7	3	3
<10>*	**mmhc**	**W-H-C**	**rbl**	**iwl**	**7.2**	**9.0**	**12.2**	**3**	**15**
<11>	rsmax2	W-H-C	ibl+rbl	-	8.9	11.1	15.1	4	3
<12>	**rsmax2**	**W-H-C**	**rbl**	**iwl**	**7.2**	**9.0**	**12.2**	**3**	**15**
<13>	mmhc	B-A-C	ibl+rbl	-	10.9	14.2	19.0	2	4
<14>	mmhc	B-A-C	rbl	iwl	11.1	14.8	19.6	1	15
<15>	rsmax2	B-A-C	ibl+rbl	-	8.9	10.5	14.7	5	4
<16>	rsmax2	B-A-C	rbl	iwl	11.1	14.8	19.6	1	15
<17>	mmhc	B-C	ibl+rbl	-	10.9	14.2	19.0	2	4
<18>	mmhc	B-C	rbl	iwl	11.1	14.8	19.6	1	15
<19>	rsmax2	B-C	ibl+rbl	-	9.1	11.0	15.2	3	4
<20>	rsmax2	B-C	rbl	iwl	11.1	14.8	19.6	1	15
<21>	mmhc	W-H-A	ibl+rbl	-	10.8	14.0	18.8	2	2
<22>	mmhc	W-H-A	rbl	iwl	8.1	10.2	13.9	2	15
<23>	rsmax2	W-H-A	ibl+rbl	-	9.8	12.0	16.5	4	3
<24>	rsmax2	W-H-A	rbl	iwl	8.1	10.2	13.9	2	15
saturated	-	A-H-W-C-B	-	-	5.3	8.3	10.4	45	36

Table 8.2
The 32 BNs we explored, one for each line. The columns are the learning algorithm (Algo.), the set of covariates (Cova.), the blacklist (BL, if any), the whitelist (WL, if any). Their performance has been assessed with the bias (|Bias|), the standard deviation (SD), the SEP score, the number of arcs going from a covariate to a response variable (c2v) and the number of arcs going from a response variable to another response variable (v2v). "rbl" is the list of arcs from a variable to a covariate. The selected models are shown in bold and the model chosen for interpretation is marked with a "*".

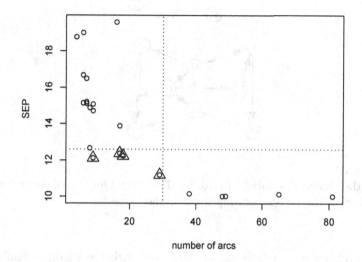

number of arcs

Figure 8.11
Each of the 32 models in Table 8.2 is shown according to the total number of arcs (excluding those linking two covariates) and to the SEP score. The models with both a small SEP and small number of arcs are surrounded with a triangle (some are superimposed); they are marked in bold in Table 8.2.

Table 8.2 summarises the BNs we created previously and those resulting from a systematic search with 6 subsets of covariates (including the complete set) and with two algorithms (mmhc and rsmax2, called with their default arguments), using either the whitelist or the blacklist above. rsmax2 is implemented using the Semi-Interleaved Hiton-PC algorithm (si.hiton.pc) in the *restrict* phase and hill-climbing (hc) in the *maximise* phase. The quality of each of these BNs is assessed with the three scores and the number of arcs either between a covariate and a variable or between two variables.

The correlation between the bias and the standard deviation over all the BNs (except of the saturated one) is very high (> 0.99) and suggests that both criteria are equivalent to each other and to the global SEP.

To get a synthetic view of the quality of the different models, we plotted the respective SEP scores against the total number of arcs in Figure 8.11. A number of BNs are highlighted with triangles and appear to reach a good balance between complexity and SEP: they are those shown in bold in Table 8.2. Some of these BNs, while highlighted, are not interesting as candidate models for prediction because they contain isolated nodes (that is, nodes that are completely disconnected from the rest of the network). Perhaps

W–H–C / W–List / mmhc

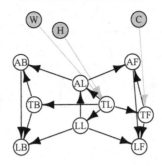

Figure 8.12

DAG of the chosen BN, labelled <10> in Table 8.2. Only three covariates are included, with three arcs pointing towards the variables.

a sensible choice is to retain one of the GBNs with the smallest number of covariates, denoted <10> in Table 8.2: its structure is shown in Figure 8.12. Equivalently, we could choose <12> which has the same structure as <10>. The saturated BN provides good predictive accuracy but it is the worst for interpretation, and it is unsurprising to note that it was outperformed by some BNs with a dramatically simple structure for the SEP criterion.

The interpretation of the chosen BN is straightforward. Height and weight have a direct influence on the lean mass of the trunk (TL), and in turn that influences the two other lean segments (LL and AL) which are all adjacent to each other. The same is true for the other two components: TF, LF and AF; and TB, LB and AB. Component-regions are directly linked within the component, the trunk being the pivot. Finally, an important correction is provided by the waist circumference (C) on the fat of the trunk (TF), which has a knock-on effect on the two other fat nodes.

The proportion of variance explained for each variable by its parents provides a local quantitative assessment of the BN's quality. To compute it we can use the sum of squares from the corresponding analysis of variance, for instance on the variable TL:

```
av.tl <- anova(lm(TL ~ H + W, data = dval))
1 - av.tl["Residuals", "Sum Sq"] / sum(av.tl[, "Sum Sq"])
[1] 0.876
```

Table 8.3 reports these proportions for all variables. They are quite good, except for TB (0.50). Note, however, that there is some arbitrariness in the choice of the parents because different equivalent DAGs could have been used; we rely on an expert's choice. Furthermore, it is worth pointing out that these proportions do not measure the quality of predictions: they only assess the

	Parent(s)	Variance explained
TF	TL + C	0.94
LF	TF + AF + LL	0.74
AF	TF + AL	0.80
TL	W + H	0.81
LL	TL	0.77
AL	TL + LL	0.77
TB	TL	0.50
LB	LL + TB + AB	0.88
AB	AL + TB	0.78

Table 8.3
The proportion of variance explained for each of the nine response variables by their respective parents within the chosen BN.

strength of the arcs included in the BN. The two coincide only for nodes in which all parents are covariates, as is the case for TL.

8.3 Further Reading

There are few books in the literature that describe real-world applications of BNs, as we did in this chapter. For the interested reader, we suggest Pourret et al. (2008) and Nagarajan et al. (2013); Chapters 5 and 11 in Korb and Nicholson (2011); and Chapters 6–8 in Holmes and Jaim (2008). The data analysed therein span several different fields, including medical diagnosis, genetics, education, forensic, finance and industrial systems. A collection of applications specific to genetics and genomics is available in Sinoquet and Mourad (2013).

A

Graph Theory

A.1 Graphs, Nodes and Arcs

A graph $G = (\mathbf{V}, A)$ consists of a non-empty set $\mathbf{V}$ of *nodes* or *vertices* and a finite (but possibly empty) set A of pairs of vertices called *arcs*, *links* or *edges*.

Each arc $a = (u, v)$ can be defined either as an ordered or an unordered pair of nodes, which are said to be *connected* by and *incident* on the arc and to be *adjacent* to each other. Here we will restrict ourselves to graphs having zero or one connection between any pair of nodes. Since they are adjacent, u and v are also said to be *neighbours*. If (u, v) is an ordered pair, u is said to be the *tail* of the arc and v the *head*; then the arc is said to be *directed* from u to v, and is usually represented with an arrowhead in v $(u \to v)$. It is also said that the arc *leaves* or is *outgoing* for u, and that it *enters* or is *incoming* for v. If (u, v) is unordered, u and v are simply said to be incident on the arc without any further distinction. In this case, they are commonly referred to as *undirected arcs* or *edges*, denoted with $e \in E$ and represented with a simple line $(u - v)$.

The characterisation of arcs as directed or undirected induces an equivalent characterisation of the graphs themselves, which are said to be *directed graphs* (denoted with $G = (\mathbf{V}, A)$) if all arcs are directed, *undirected graphs* (denoted with $G = (\mathbf{V}, E)$) if all arcs are undirected and *partially directed* or *mixed graphs* (denoted with $G = (\mathbf{V}, A, E)$) if they contain both arcs and edges.

An example of each class of graphs is shown in Figure A.1. The first graph (on the left) is an undirected graph, in which:

- the node set is $\mathbf{V} = \{A, B, C, D, E\}$ and the edge set is $E = \{$ $(A - B)$, $(A - C)$, $(A - D)$, $(B - D)$, $(C - E)$, $(D - E)$ $\}$;

- arcs are undirected, so *i.e.*, $A - B$ and $B - A$ are equivalent and identify the same edge;

- likewise, A is connected to B, B is connected to A, and A and B are adjacent.

The second graph (centre) is a directed graph. Unlike the undirected graph we just considered:

- the directed graph is characterised by the arc set $A = \{(A \to B), (C \to A),$

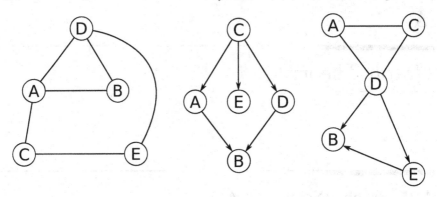

Figure A.1
An undirected graph (left), a directed graph (centre) and a partially directed
graph (right).

$(D \rightarrow B)$, $(C \rightarrow D)$, $(C \rightarrow E)\}$ instead of an edge set E, even though the
node set **V** is the same as before;

- arcs are directed, so *i.e.*, $A \rightarrow B$ and $B \rightarrow A$ identify different arcs; $A \rightarrow B$
  $\in A$ while $B \rightarrow A \notin A$. Furthermore, it is not possible for both arcs to be
  present in the graph because there can be at most one arc between each pair
  of nodes;

- again, A and B are adjacent, as there is an arc $(A \rightarrow B)$ from A to B. $A \rightarrow B$
  is an outgoing arc for A (the tail), an incoming arc for B (the head) and an
  incident arc for both A and B.

The third graph (on the right) is a mixed graph, which is characterised by the
combination of an edge set $E = \{(A - C), (A - D), (C - D)\}$ and an arc set
$A = \{(D \rightarrow E), (D \rightarrow B), (E \rightarrow B)\}$.

  An undirected graph can always be constructed from a directed or partially
directed one by substituting all the directed arcs with undirected ones; such a
graph is called the *skeleton* or the *underlying undirected graph* of the original
graph.

A.2   The Structure of a Graph

The pattern with which the arcs appear in a graph is referred to as either
the *structure* of the graph or the *configuration* of the arcs. In the context of
this book it is assumed that the vertices u and v incident on each arc are
distinct (if $u = v$ the arc is called a *loop*) and that there is at most one

arc between them (so that (u, v) uniquely identifies an arc). The simplest structure is the *empty graph*, that is the graph with no arcs; at the other end of the spectrum are *saturated graphs*, in which each node is connected to every other node. Graphs between these two extremes are said to be *sparse* if they have few arcs compared to the number of nodes, and *dense* otherwise. While the distinction between these two classes of graphs is rather vague, a graph is usually considered sparse if $O(|E| + |A|) = O(|\mathbf{V}|)$.

The structure of a graph determines its properties. Some of the most important deal with *paths*, sequences of arcs or edges *connecting* two nodes, called *end-vertices* or *end-nodes*. Paths are directed for directed graphs and undirected for undirected graphs, and are denoted with the sequence of vertices $(v_1, v_2, \ldots, v_n)$ incident on those arcs. The arcs connecting the vertices $v_1, v_2, \ldots, v_n$ are assumed to be unique, so that a path passes through each arc only once. In directed graphs it is also assumed that all the arcs in a path follow the same direction, and we say that a path *leads from* v_1 (the tail of the first arc in the path) *to* v_n (the head of the last arc in the path). In undirected and mixed graphs (and in general when referring to a graph regardless of which class it belongs to), arcs in a path can point in either direction or be undirected. Paths in which $v_1 = v_n$ are called *cycles*, and are treated with particular care in BN theory.

The structure of a directed graph defines a partial ordering of the nodes if the graph is *acyclic*, that is, if it does not contain any cycle or loop. In that case the graph is called a *directed acyclic graph* (DAG). This ordering is called an *acyclic* or *topological ordering* and is induced by the direction of the arcs. It must satisfy the following property:

$$\{\exists \text{ a path from } v_i \text{ to } v_j\} \Rightarrow \{v_i \prec v_j\} \tag{A.1}$$

so when a node v_i precedes v_j there can be no path from v_j to v_i. According to this definition the first nodes are the *root nodes*, which have no incoming arcs, and the last are the *leaf nodes*, which have no outgoing arcs. Furthermore, if there is a path leading from v_i to v_j, v_i precedes v_j in the sequence of the ordered nodes. In this case v_i is called an *ancestor* of v_j and v_j is called a *descendant* of v_i. If the path is composed by a single arc by analogy x_i is a *parent* of v_j and v_j is a *child* of v_i.

Consider, for instance, node A in the DAG shown in Figure A.2. Its neighbourhood is the union of the parents and children; adjacent nodes necessarily fall into one of these two categories. Its parents are also ancestors, as they necessarily precede A in the topological ordering. Likewise, children are also descendants. Two examples of topological ordering induced by the graph structure are

$$(\{F, G, H\}, \{C, B\}, \{A\}, \{D, E\}, \{L, K\}) \tag{A.2}$$

and

$$(\{F\}, \{B\}, \{G\}, \{C\}, \{A, H\}, \{E\}, \{L\}, \{D\}, \{K\}) . \tag{A.3}$$

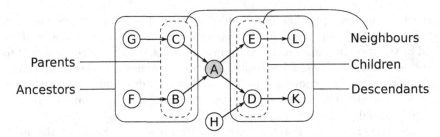

Figure A.2
Parents, children, ancestors, descendants and neighbours of node A in a directed graph.

Indeed, the nodes are only *partially ordered*; for example, no ordering can be established among root nodes or leaf nodes. In the examples above, the nodes enclosed in each set of curly brackets can be permuted in any possible way and still satisfy Equation (A.1) for the DAG in Figure A.2. As a result, in practice the topological ordering of a DAG is defined over a set of unordered sets of nodes, denoted with $V_i = \{v_{i_1}, \ldots, v_{i_k}\}$, defining a partition of $\mathbf{V}$.

A.3   Further Reading

For a broader coverage of the properties of directed and mixed graphs, we refer the reader to the monograph by Bang-Jensen and Gutin (2009), which at the time of this writing is the most complete reference on the subject. For undirected graphs, we refer to the classic book of Diestel (2005).

B

Probability Distributions

B.1 General Features

A *probability distribution* is a function that assigns a *probability* to each measurable subset of a set of *events*. It is associated with a *random variable*, here denoted as X.

A *discrete probability distribution* can assume only a finite or countably infinite number of values $\mathbf{U}$, such as

$$\{A, B, C, D\} \quad \text{or} \quad \{[0,5], (5,7], (8,10]\} \quad \text{or} \quad n \in \mathbb{N}, \qquad (B.1)$$

and is characterised by a probability function $\Pr(\cdot)$ that satisfies

$$\sum_{u \in \mathbf{U}} \Pr(X = u) = 1 \quad \text{and} \quad \Pr(X = u) \in [0,1] \quad \text{for all } u. \qquad (B.2)$$

A *continuous probability distribution* must assume an infinite number of values $\mathbf{U}$, typically $\mathbb{R}$ or an interval of real numbers, and is characterised by a *density function* $f(\cdot)$ that satisfies

$$\int_{\mathbf{U}} f(X = u) du = 1 \quad \text{and} \quad f(X = u) \geqslant 0 \quad \text{for all } u. \qquad (B.3)$$

$\Pr(u)$ and $f(u)$ are often used as shorthand notations when it is clear which variable we are referring to, along with more precise notations such as $\Pr_X(u)$ and $f_X(u)$. In addition, sometimes the distinction between discrete and continuous distributions is unimportant and we use $\Pr(\cdot)$ instead of $f(\cdot)$. Note that $\Pr(u) \leqslant 1$, but $f(u)$ has no such restriction and can be an arbitrarily large positive number. Furthermore, note that points have null measure, and thus zero probability, for all continuous distributions; intervals can have positive probability. We say that X *follows* or *is distributed* as a certain probability distribution, and we denote it with $X \sim \Pr(\cdot)$ or $X \sim f(\cdot)$.

Summing $\Pr(\cdot)$ over $\mathbf{U}$ or some subset of it produces many quantities that are extremely useful to study distributions when $\mathbf{U}$ admits an ordering. For instance, this is the case for discrete random variables defined over $\mathbb{N}$ and continuous ones. Some examples are:

- The *cumulative distribution function* (CDF),

$$F(u^*) = \sum_{u \leqslant u^*} \Pr(u) \qquad \text{or} \qquad F(u^*) = \int_{-\infty}^{u^*} \Pr(u)du, \qquad \text{(B.4)}$$

which is the probability of getting value no greater than u^*. The quantile function is just the inverse of $F(\cdot)$:

$$Q(p) = u \iff F(u) = p. \qquad \text{(B.5)}$$

$F(\cdot)$ is usually estimated from an observed sample $x_1, \ldots, x_n$ with the *empirical cumulative distribution function* (ECDF),

$$\widehat{F}(u^*) = \frac{1}{n} \sum_{i=1}^{n} \mathbb{1}(x_i \leqslant u^*), \qquad \text{(B.6)}$$

the proportion of the observations that have a value no greater than u^*.

- The *mean*, *expected value* or *expectation*,

$$E(X) = \sum_{u \in U} u \Pr(u) \qquad \text{or} \qquad E(X) = \int_{U} u \Pr(u)du, \qquad \text{(B.7)}$$

usually estimated with the *empirical mean* $\overline{x} = \frac{1}{n} \sum_{i=1}^{n} x_i$. Note that for some distributions $E(X)$ may not be finite.

- The *variance*,

$$\text{VAR}(X) = E\left[(X - E(X))^2\right] = E(X^2) - E(X)^2 \qquad \text{(B.8)}$$

which measures how much the distribution of X is spread around $E(X)$. Its positive square root is called the *standard deviation* of X. The variance is often estimated as $\frac{1}{n} \sum_{i=1}^{n} (x_i - \overline{x})^2$. Again, for some distributions $\text{VAR}(X)$ may not be finite.

Both mean and variance are easy to compute for *linear transformations* of X, that is, when adding and multiplying by some constants $a, b \in \mathbb{R}$:

$$E(aX + b) = a E(X) + b \qquad \text{and} \qquad \text{VAR}(aX + b) = a^2 \text{VAR}(X). \qquad \text{(B.9)}$$

B.2 Marginal and Conditional Distributions

All the definitions above implicitly assume that X takes values in a one-dimensional space U such as $\mathbb{N}$ or $\mathbb{R}$; in that case X is called a *univariate random variable.* On the other hand, a *multivariate random variable* $\mathbf{X}$ takes values in a multi-dimensional space such as $\mathbb{R}^k, k \geqslant 2$. It is also called a *random vector* because it can be characterised by the *joint distribution* of a vector of univariate variables,

$$\mathbf{X} = [X_1 \, X_2 \, \cdots \, X_k], \tag{B.10}$$

one for each dimension of $\mathbf{X}$. Therefore, integrals and summations can be used as before but have to be defined over all the k dimensions. For example, the mean of $\mathbf{X}$ is a vector of length k,

$$\mathrm{E}(\mathbf{X}) = [\mathrm{E}(X_1) \, \mathrm{E}(X_2) \, \cdots \, \mathrm{E}(X_k)], \tag{B.11}$$

and the variance is replaced by a $k \times k$ *covariance matrix*

$$\mathrm{COV}(\mathbf{X}) = \begin{bmatrix} \mathrm{VAR}(X_1) & \mathrm{COV}(X_1, X_2) & \cdots & \mathrm{COV}(X_1, X_k) \\ \mathrm{COV}(X_2, X_1) & \mathrm{VAR}(X_2) & \cdots & \mathrm{COV}(X_2, X_k) \\ \vdots & \vdots & \ddots & \vdots \\ \mathrm{COV}(X_k, X_1) & \mathrm{COV}(X_k, X_2) & \cdots & \mathrm{VAR}(X_k) \end{bmatrix} \tag{B.12}$$

where $\mathrm{COV}(X_i, X_j) = \mathrm{E}(X_i X_j) - \mathrm{E}(X_i) \mathrm{E}(X_j)$ is the *covariance* between X_i and X_j; note that $\mathrm{COV}(X_i, X_i) = \mathrm{VAR}(X_i)$ and that $\mathrm{COV}(X_i, X_j) = \mathrm{COV}(X_j, X_i)$. The empirical estimate of $\mathrm{COV}(X_i, X_j)$ is $\frac{1}{n} \sum_{l=1}^{n} (x_{li} - \overline{x}_i)(x_{lj} - \overline{x}_j)$, where x_{li} is the lth observation for X_i and x_{lj} is the lth observation for X_j. A related quantity which is easier to interpret is *correlation*, defined as

$$\mathrm{COR}(X_i, X_j) = \frac{\mathrm{COV}(X_i, X_j)}{\sqrt{\mathrm{VAR}(X_i)} \sqrt{\mathrm{VAR}(X_j)}} \tag{B.13}$$

and estimated using the empirical estimates of variance and covariance.

The univariate linear transformation in Equation (B.9) can be reformulated using a $h \times 1$ vector $\mathbf{b}$ and a $h \times k$ matrix A:

$$\mathrm{E}(A\mathbf{X} + \mathbf{b}) = A\,\mathrm{E}(\mathbf{X}) + \mathbf{b} \quad \text{and} \quad \mathrm{COV}(A\mathbf{X} + \mathbf{b}) = A\,\mathrm{COV}(\mathbf{X})A^T. \tag{B.14}$$

The distribution of each X_i in $\mathbf{X}$ can be considered without taking into account its relationship with other variables; in this case it is called the *marginal distribution* of X_i. We can derive it from the distribution of $\mathbf{X}$ by summing or integrating over all the possible values $\mathbf{U}_j, j \neq i$ of all $X_j, j \neq i$:

$$\mathrm{Pr}(X_i) = \sum_{u_1 \in \mathbf{U}_1} \cdots \sum_{u_{i-1} \in \mathbf{U}_{i-1}} \sum_{u_{i+1} \in \mathbf{U}_{i+1}} \cdots \sum_{u_k \in \mathbf{U}_k} \mathrm{Pr}(\mathbf{X}), \tag{B.15}$$

$$\mathrm{Pr}(X_i) = \int_{\mathbf{U}_1} \cdots \int_{\mathbf{U}_{i-1}} \int_{\mathbf{U}_{i+1}} \cdots \int_{\mathbf{U}_k} \mathrm{Pr}(\mathbf{X}) d\mathbf{X}. \tag{B.16}$$

Extending these definitions to obtain the marginal distribution of more than one variable is trivial; summation (or integration) is carried out as above, just on all the other variables in $\mathbf{X}$.

Another distribution of interest for X_i is its *conditional distribution*, that is, the distribution of X_i when the values of some other variables $X_{j_1}, \ldots, X_{j_m}$ in $\mathbf{X}$ are fixed to specific values. It can be derived as follows:

$$\Pr(X_i = u_i \mid X_{j_1} = u_{j_1}, \ldots, X_{j_m} = u_{j_m}) = \frac{\Pr(X_i = u_i, X_{j_1} = u_{j_1}, \ldots, X_{j_m} = u_{j_m})}{\Pr(X_{j_1} = u_{j_1}, \ldots, X_{j_m} = u_{j_m})}. \quad (B.17)$$

Again, extending this definition to include more than one variable just requires adding the corresponding terms in the probability and density functions.

B.3   Discrete Distributions

Discrete distributions in common use for BNs are the *binomial* and the *multinomial* distributions and we will focus on them. Nevertheless, as shown in Chapter 3, any distribution can be used in the general case so we will also mention others that are important in practical situations.

B.3.1   Binomial Distribution

The binomial distribution models the number of successes in a sequence of n independent experiments with two possible outcomes; each experiment yields a positive outcome (often called a *success*) with identical probability p. It is usually denoted with $Bi(n, p)$ and has probability function

$$\Pr(X = x) = \binom{n}{x} p^x (1 - p)^{(n-x)}, \quad n \in \mathbb{N}, x \in \{0, \ldots n\}, p \in [0, 1]. \quad (B.18)$$

When $p = 0$ (or $p = 1$) the distribution is degenerate in the sense that X can take only the value 0 (or n). Expectation and variance are easily calculated:

$$\mathrm{E}(X) = np \qquad \text{and} \qquad \mathrm{VAR}(X) = np(1 - p). \qquad (B.19)$$

So, for n fixed, the variance is maximal when $p = \frac{1}{2}$ and tends to zero when p tends to 0 or 1.

The maximum likelihood estimate of p is simply $\hat{p} = \frac{r}{n}$, the number r of successes over the number n of experiments. Such an estimate is problematic when $r = 0$ or $r = n$, as discussed in Section 1.4.

B.3.2 Multinomial Distribution

The multinomial distribution is the multivariate extension of the binomial, which is required when there are three or more possible outcomes instead of two. Strictly speaking, the binomial is the multivariate distribution of the frequencies of the two outcomes; but as the frequency of one outcome is uniquely determined from the other as $(n - X)$, it is redundant to make it explicit in the notation. Along with the binomial, it's called a *categorical* distribution if the outcomes do not have any explicit ordering.

Given a sequence of n independent trials each having identical probabilities $\mathbf{p} = (p_1, \ldots, p_k)$ for k possible outcomes, the vector of the associated counts $\mathbf{X} = (X_1, \ldots, X_k)$ is said to follow a multinomial distribution and it is denoted as $Mu(n, \mathbf{p})$. The probability function is

$$\Pr(\mathbf{X} = \mathbf{x}) = \frac{n!}{x_1! x_2! \cdots x_k!} p_1^{x_1} p_2^{x_2} \cdots p_k^{x_k}, \quad n \in \mathbb{N}, \sum_i x_i = n, \sum_i p_i = 1.$$
(B.20)

It is important to note that even though each X_i is a random variable following a $Bi(n, p_i)$, $X_1, \ldots, X_k$ are linked by the simple fact to sum to n. As a consequence, they are negatively correlated $(\mathrm{COR}(X_i, X_j) < 0)$ and the covariance matrix is not full rank since $\mathrm{VAR}(\sum_i X_i) = 0$. Furthermore,

$$E(\mathbf{X}) = n\mathbf{p} \quad \text{and} \quad \mathrm{VAR}(\mathbf{X}) = n \left(\mathrm{diag}(\mathbf{p}) - \mathbf{p}\mathbf{p}^T \right), \quad (B.21)$$

where $\mathrm{diag}(\mathbf{p})$ denotes the diagonal matrix with vector $\mathbf{p}$ on its diagonal.

B.3.3 Other Common Distributions

B.3.3.1 Bernoulli Distribution

The *Bernoulli distribution*, $Ber(p)$ is a particular case of $Bi(n, p)$ when $n = 1$.

B.3.3.2 Poisson Distribution

The *Poisson* distribution can be derived from the binomial distribution. Consider a rare animal living in a particular region. If we partition that region into N land plots, when N is large enough the probability that two or more animals are present in any plot is negligible. While the number of animals can be modelled as a $Bi(N, p)$ where the probability p depends on the size of the plots, it is logical to assume that in fact $p = \frac{\lambda}{N}$.

It occurs that when $N \to \infty$,

$$Bi\left(N, \frac{\lambda}{N}\right) \to Pois(\lambda), \quad (B.22)$$

where $Pois(\lambda)$ is a Poisson distribution, and λ is in fact the density of presence in the region. The probability function of $Pois(\lambda)$ is

$$\Pr(X = x) = \frac{\lambda^x e^{-\lambda}}{x!}, \qquad\qquad x \in \mathbb{N}, \lambda \geqslant 0, \qquad (\text{B.23})$$

and both mean and variance are equal to λ:

$$\mathrm{E}(X) = \lambda \qquad\qquad \text{and} \qquad\qquad \mathrm{VAR}(X) = \lambda. \qquad (\text{B.24})$$

B.4   Continuous Distributions

The *normal* or *Gaussian distribution* plays a central role among continuous distributions. Like the binomial distribution, it can be conveniently extended into a multivariate distribution. Some indications about the estimation of the parameters of the normal distribution are provided in Section B.2. In addition, we also consider the *beta* distribution (and its multivariate version) for its close relationship with the binomial (multinomial) discrete distribution.

B.4.1   Normal Distribution

A normal or Gaussian distribution has density

$$\Pr(x; \mu, \sigma^2) = \frac{1}{\sqrt{2\pi\sigma^2}} \exp\left\{-\frac{1}{2\sigma^2}(x - \mu)^2\right\}, \quad x, \mu \in \mathbb{R}, \sigma^2 > 0. \quad (\text{B.25})$$

It is denoted with $N(\mu, \sigma^2)$. Some simple calculations show that the two parameters μ and σ^2 have a very direct interpretation, since

$$\mathrm{E}(X) = \mu \qquad\qquad \text{and} \qquad\qquad \mathrm{VAR}(X) = \sigma^2. \qquad (\text{B.26})$$

When $\mu = 0$ and $\sigma^2 = 1$, the associated random variable follows a *standard normal distribution*, and it is straightforward to check that $\frac{X-\mu}{\sigma} \sim N(0,1)$. It is also worth noting that even if X is defined over the whole $\mathbb{R}$ and any value between $-\infty$ and $+\infty$ has a strictly positive density,

$$\Pr\left(X \notin [\mu - 4\sigma, \mu + 4\sigma]\right) < 5.10^{-5}. \qquad (\text{B.27})$$

B.4.2   Multivariate Normal Distribution

The multivariate extension of the normal distribution is called multivariate normal, multinormal or multivariate Gaussian distribution. Its density function is

$$\Pr(\mathbf{x}; \boldsymbol{\mu}, \Sigma) = \frac{1}{\sqrt{2\pi \det(\Sigma)}} \exp\left\{-\frac{1}{2}(\mathbf{x} - \boldsymbol{\mu})^T \Sigma^{-1}(\mathbf{x} - \boldsymbol{\mu})\right\}, \quad \mathbf{x}, \boldsymbol{\mu} \in \mathbb{R}^k,$$

$$(\text{B.28})$$

where μ is a $k \times 1$ mean vector and Σ is a $k \times k$ positive-semidefinite covariance matrix. The distribution is denoted with $N_k(\mu, \Sigma)$.

A multivariate normal random variable has two very convenient properties:

1. An affine transformation of a normal vector is still a normal vector. More precisely, supposing that $\mathbf{b}$ is a vector of size h and A an $h \times k$ matrix:

$$\mathbf{X} \sim N_k(\mu, \Sigma) \quad \Longrightarrow \quad A\mathbf{X} + \mathbf{b} \sim N_h\left(\mathbf{b} + A\mu, A\Sigma A^T\right). \quad (\text{B.29})$$

2. The conditional distribution for a subset A of components of $\mathbf{X}$ given the other components B of a multivariate normal is again a multivariate normal distribution with

$$\mu_{A|B} = \mu_A + \Sigma_{AB}\Sigma_{BB}^{-1}(\mathbf{x}_b - \mu_B)$$
$$\Sigma_{A|B} = \Sigma_{AA} - \Sigma_{AB}\Sigma_{BB}^{-1}\Sigma_{BA} \quad (\text{B.30})$$

where

$$\mu = \begin{bmatrix} \mu_A \\ \mu_B \end{bmatrix} \quad \text{and} \quad \Sigma = \begin{bmatrix} \Sigma_{AA} & \Sigma_{AB} \\ \Sigma_{BA} & \Sigma_{BB} \end{bmatrix}. \quad (\text{B.31})$$

Note that Equation (B.30) implicitly assumes that Σ_{BB} is full rank; that is, no redundancy occurs in the conditioning variables which are linearly independent. Theoretically, this is not an insurmountable problem since redundant variables can be removed without changing the nature of the problem. However, from a practical point of view it is often a difficult task since the assessment of the rank of a matrix (even a non-negative one) is a non-trivial numerical problem.

B.4.3   Other Common Distributions

Two important distributions related to the univariate normal distribution are *Student's t* and the *Chi-square* (χ^2) distribution. *Beta* and *Dirichlet* distributions have been included, because they are conjugate distributions (see Section B.5) of the binomial and the multinomial.

B.4.3.1   *Chi-Square Distribution*

The χ^2 distribution can be viewed as the sum of the squares of ν independent standard normal random variables. ν is then the only parameter, called its *degrees of freedom*; the distribution is denoted with χ_ν^2. Accordingly, the sum of two independent χ^2 random variables with ν_1 and ν_2 degrees of freedom follows a χ^2 distribution with $\nu_1 + \nu_2$ degrees of freedom. The expectation and variance are :

$$\mathrm{E}(X) = \nu \quad \text{and} \quad \mathrm{VAR}(X) = 2\nu. \quad (\text{B.32})$$

This distribution can be generalised by introducing a non-centrality parameter induced by non-centred generative normal variables of unit variance. More precisely, let

$$X = \sum_{i=1}^{\nu} U_i^2 \quad \text{where the } U_i \text{ are independent } N(\mu_i, 1); \qquad (B.33)$$

then $X \sim nc\chi^2(\lambda)$ with $\lambda = \sum_{i=1}^{\nu}(\mu_i^2)$. Equation (B.32) generalises with

$$\mathrm{E}(X) = \nu + \lambda, \qquad (B.34)$$
$$\mathrm{VAR}(X) = 2(\nu + 2\lambda), \qquad (B.35)$$
$$\mathrm{E}\left([X - E(X)]^3\right) = 8(\nu + 3\lambda). \qquad (B.36)$$

It can also be shown that the χ^2 distribution is a particular case of the *gamma* distribution, allowing the ν parameter to take non-integer but positive values.

B.4.3.2 *Student's t Distribution*

The need for the Student's t distribution arises when standardising a normal random variable with an estimate of its variance. It can be introduced as the ratio of a standard normal variable with the square root of an independent χ^2_ν variable. Then, it has only one parameter, the degrees of freedom of the χ^2 variable, $\nu \in \mathbb{N}^+$. Its density function is bell-shaped and symmetric like the normal's, and differs mainly in the thickness of the tails. For high values of ν Student's t is well approximated by a $N(0,1)$ distribution, while low values result in fatter tails and assign more probability to extreme values. When $\nu > 2$, the expectation and variance are:

$$\mathrm{E}(X) = 0 \qquad \text{and} \qquad \mathrm{VAR}(X) = \frac{\nu}{\nu - 2}. \qquad (B.37)$$

As for the χ^2 distribution, a second *non-centrality* parameter can be considered to generalise the distribution.

B.4.3.3 *Beta Distribution*

The beta distribution, denoted with *Beta(a, b)*, is a common choice for a random variable restricted to the $[0, 1]$ interval. Its density function is

$$\Pr(x; a, b) = \frac{\Gamma(a + b)}{\Gamma(a)\Gamma(b)} x^{a-1}(1 - x)^{b-1}, \qquad a, b > 0, x \in [0, 1], \qquad (B.38)$$

which simplifies to a binomial if $a, b \in \mathbb{N}$ because $\Gamma(n) = (n - 1)!$ for $n \in \mathbb{N}$. In that case, a plays the role of x, b plays the role of $n - x$ and x plays the role of p in Equation (B.18). This is why it is the conjugate distribution (see Section B.5) in the beta-binomial model in Bayesian statistics. The expectation and variance are:

$$\mathrm{E}(X) = \frac{a}{a + b} \qquad \text{and} \qquad \mathrm{VAR}(X) = \frac{ab}{(a + b)^2(a + b + 1)}. \qquad (B.39)$$

This distribution is quite flexible when varying the parameters. It can easily be transposed to any finite non point interval $[a, b]$ by an affine transformation of the form

$$a + (b - a)X \qquad \text{where} \qquad a, b \in \mathbb{R}. \qquad \text{(B.40)}$$

B.4.3.4 Dirichlet Distribution

The Dirichlet distribution is a multivariate generalisation of the beta distribution in the same way as the multinomial is a generalisation of the binomial. For instance, a Dirichlet with two components is just a beta distribution, the two components summing to one. The distribution is often denoted by $Dir(\boldsymbol{a})$, $\boldsymbol{a} = (a_1, a_2, \ldots, a_k)$, and its density function is

$$\Pr(X = \mathbf{x}) = \frac{\Gamma(\sum_i a_i)}{\Gamma(a_1)\Gamma(a_2)\cdots\Gamma(a_k)} x_1^{a_1} x_2^{a_2} \cdots x_k^{a_k},$$

$$\sum_i x_i = 1, x_i \in [0, 1], a_1, \ldots, a_k > 0. \quad \text{(B.41)}$$

As was the case for the beta with the binomial, the similarity with the multinomial distribution is apparent: x_i plays the role of the p_i and a_i plays the role of the x_i in Equation (B.20). Denoting with A the sum of the parameters $\sum_i a_i$, the expectation and variance are:

$$\mathrm{E}(X) = \frac{1}{A}\boldsymbol{a} \quad \text{and} \quad \mathrm{VAR}(X) = \frac{1}{A(1 + A)}\left(\mathrm{diag}(\boldsymbol{a} - \frac{1}{A}\boldsymbol{aa}^T\right). \quad \text{(B.42)}$$

From a Bayesian point of view, the Dirichlet is the conjugate distribution (see Section B.5) for the parameter **p** of a multinomial distribution.

B.5 Conjugate Distributions

Several times in the above, we have mentioned that two distributions are conjugate. Here we will try to convey the fundamental idea of conjugacy in a few words. The concept originates in the framework of Bayesian statistics.

Consider two random variables A and B whose joint probability is defined by the marginal and conditional distributions,

$$\Pr(A, B) = \Pr(A)\Pr(B \mid A). \qquad \text{(B.43)}$$

When the conditional distribution $\Pr(A \mid B)$ belongs to the same family as $\Pr(A)$, then we say that $\Pr(A)$ is the *conjugate* of $\Pr(B \mid A)$. A simple example is the beta-binomial pair:

- let $p \sim Beta(a, b)$,

- and let $(r \mid p) \sim Bi(n, p)$,

- then $(p \mid r) \sim Beta(a + r, b + n - r)$.

The main advantage of conjugacy is to simplify to the extreme the computation of conditional distributions, since their closed form is known.

B.6   Further Reading

Literature contains many books on probability theory with varying levels of complexity. An introductory book focusing on commonly used distributions and their fundamental properties is Ross (2018). More advanced books, which cover probability theory from a more theoretical perspective, are DeGroot and Scherivsh (2012); Ash (2000); Feller (1968). A thorough derivation of the basis of probability theory as an off-shoot of measure theory is provided in Loève (1977). Furthermore, an encyclopedia about probability distributions is the series of Johnson's and Kotz's books (Kotz et al., 2000; Johnson et al., 1994, 1995, 1997, 2005).

C

A Note about Bayesian Networks

The relationship between *Bayesian networks* and *Bayesian statistics* is often unclear because of how similar their names are due to the common qualifier of *Bayesian*. Even though they are connected in many ways, it is important to underline the differences between the two.

C.1 Bayesian Networks and Bayesian Statistics

The fundamental difference originates in the term *statistics*. A statistical procedure consists in summarising the information contained in a data set, or more generally in performing some inference within the framework of some probabilistic model. A Bayesian network is just a complex, principled way of proposing a probabilistic model on a set of variables, without necessarily involving data.

The confusion originates from the following points:

- It is very convenient to present Bayesian statistics as a BN of this kind: (*Parameters*) $\mapsto$ (*Data*). The marginal distribution of the multivariate node (*Parameters*) is the prior; the conditional distribution of the multivariate node (*Data*) is the likelihood; then the posterior distribution is just the conditional distribution of the multivariate node (*Parameters*) for the observed (*Data*). In essence, a Bayesian statistical procedure can be described as a BN.

- When querying a BN as in Section 6.6.1, by fixing some nodes of the BN and updating the local probability distributions in this new context, we are in fact applying a Bayesian statistical approach, if we accept to assimilate the fixed node(s) to observation(s).

- The probabilistic model used in a statistical approach (whether Bayesian or not) can be defined through a BN. That is apparent when using Stan in Chapters 3 and 5.

- A BN is not always known in advance and many learning algorithms have been proposed to estimate them from data. This has been detailed in various parts of the book (Sections 1.5, 2.5 and 6.5).

But it is important to be clear in the fact that the probabilistic model used in a statistical approach can be defined without the help of BNs; and also that the learning of a BN from a data set can be in a Bayesian context or not.

Glossary

Here we have collected some of the technical terms used within the book. For those terms referring to graph theory, Appendix A provides a short but comprehensive introduction to the field.

acyclic graph: An acyclic graph is a graph without any cycle.

adjacent nodes: Two nodes are said to be adjacent when they are linked by an arc or an edge.

ancestor: An ancestor of a node is a node that precedes it within a directed path and therefore in the topological ordering of the graph.

arc: An arc is a directed link between two nodes, usually assumed to be distinct. Nodes linked by an arc have a direct parent-child relationship: the node on the tail of the arc is the parent node, the node on the head of the arc is the child node. A node can have no, one or several parents; a node can have no, one or several children. Sometimes the expression "undirected arcs" is used to refer to edges, *i.e.*, undirected links.

Bayes, Thomas: Bayes was an English Presbyterian minister (1701–1761). He wrote some notes about deducing causes from effects, published after his death by a friend of his under the title *"An Essay towards solving a Problem in the Doctrine of Chances"*. They contain the original formulation of the conditional probability theorem, which is known as *Bayes' theorem*. This important result was independently published in a better mathematical form by Pierre Simon Laplace.

Bayesian networks (BNs): A Bayesian network defines a joint probability distribution over a set of variables and the corresponding local univariate distributions. The DAG associated with the BN determines whether each of them is marginal or conditional on other variables. Each node is associated with one variable in the BN. Root nodes are given marginal distributions; other nodes have distributions conditional only on the values of the respective parents. A formal link between the conditional independencies in the BN and a DAG is given in Definition 6.1. The term *Bayesian network* comes from the recursive use of Bayes' theorem to decompose the joint distribution into the individual distributions of the nodes and other distributions of interest, following the dependence structure given by the DAG.

child node: See the definition for *arc*.

clique: In an undirected graph, a clique is a subset of nodes such that every two nodes in the subset are adjacent. A clique is said to be maximal when including another node in the clique means it is no longer a clique.

conditional distribution: The conditional probability distribution of a random variable A with respect to another random variable B is its probability distribution when B is known or restricted, particularly to a given value b. Generally, this distribution depends on b and it is denoted by $\Pr(A \mid B = b)$. When A and B are independent, $\Pr(A \mid B = b) = \Pr(A)$ for any b, that is, the conditional distribution is identical to the marginal distribution and it does not depend on b.

configuration: The configuration of the arcs or edges of a graph is synonymous with the structure of a graph.

connected: Two nodes are said to be connected when there exists a path linking them; when the path is of length one (*e.g.*, a single arc or edge) they are adjacent.

connected graph: A connected graph is an undirected graph in which every pair of nodes is connected by a path; or a directed graph whose skeleton is connected (*i.e.*, disregarding the direction of the arcs and substituting them with edges).

convergent connection: A convergent connection is one of the three fundamental connections between three nodes (see Figure 1.3 on page 21). Two nodes are parents of the third: $A \rightarrow B \leftarrow C$.

CPDAG: Acronym for Completed Partially Directed Acyclic Graph. The CPDAG of a (partially) directed acyclic graph is the partially directed graph built on the same set of nodes, keeping the same v-structures and the same skeleton, completed with the compelled arcs. Two DAGs having the same CPDAG are equivalent in the sense that they result in BNs describing the same probability distribution.

cycle: A cycle is a path having identical end-nodes, that is, starting from and ending on the same node. Note that it is possible to have cycles in both directed and undirected graphs.

d-separation: Two subsets of variables, say **A** and **B**, are said to be d-separated in a BN with respect to a third, say **C**, when they are graphically separated according to the conditions in Definition 6.2. Following Definition 6.3, this also means that any variable in **A** is independent from any variable in **B** conditional on the variables in **C**.

DAGs: Acronym for Directed Acyclic Graph. It is a directed graph with no cycles. Sometimes, there is some confusion between BNs and DAGs; a DAG expresses only the conditional independence structure of a BN. Nevertheless, the DAG representation is very useful to discuss the construction or the interpretation of a BN.

descendant: A descendant of a node is a second node following it in a directed path.

directed edge: Synonymous with arc.

directed graph: A directed graph is a graph where all links are arcs (*i.e.*, have a direction).

divergent connection: A divergent connection is one of the three fundamental connections between three nodes (see Figure 1.3 on page 21). A parent with two children: $A \leftarrow B \rightarrow C$.

edge: An edge is an undirected link between two nodes. Sometimes the expression "directed edge" is used to refer to arcs.

empty graph: An empty graph is a graph with no edges or arcs at all, the node set is commonly assumed to be non-empty.

end-node: An end-node is either the first or the last node of a given path.

graph: A graph consists in a non-empty set of nodes where each pair of nodes is linked by zero or one link. Links can be arcs or edges; and then the graph is said to be directed (only arcs), undirected (only edges) or partially directed (both arcs and edges).

independence: In a probabilistic framework two random variables are said to be independent if and only if the conditional probability distribution of each of them with respect to the other is identical to the marginal distribution. It is a symmetric property. At an intuitive level, this means that knowing the value of one variable does not modify our knowledge about the distribution of the other.

joint distribution: The joint probability distribution of a set of random variables provides the probability of any combination of events over them. If the variables are independent the joint distribution simplifies into the product of the marginal distributions.

leaf node: A leaf node is a node in a DAG without any child.

link: An arc or an edge connecting two nodes.

marginal distribution: The marginal probability distribution of a random variable is its probability distribution, defined without taking into account other random variables that may be related. The knowledge of the marginal distributions of two variables is not sufficient to deduce their joint distribution without further assumptions, such as that they are independent.

On the contrary, it is always possible to deduce marginal distributions from the joint distribution. Marginal distributions can also be derived for subsets of random variables.

mixed graphs: Synonymous with partially directed graphs.

neighbour-nodes: The neighbours of a node are all the nodes that are adjacent to it.

nodes: The nodes, together with the arcs, are the fundamental components of DAGs; and in the context of BNs they also refer to the random variables in the model.

parent node: See the definition for *arc*.

partially directed: See the definition of *graph*.

path: A path is a set of successive links. It is usually described by an ordered sequence of linked nodes. When the path comprises some arcs, all of them must follow the same direction; then it is a *directed path*.

probability distribution: The probability distribution of a random variable is a function defining the probability of each possible event described by the random variable. A probability is a real number in $[0, 1]$. A probability of zero means that the event is impossible; a probability of one means that the event is certain. Even though probability distributions have different mathematical forms for discrete and continuous variables, they interact in similar ways. When the random variable involves several components, the *joint distribution* refers to events describing all the components; a *marginal distribution* refers to events describing a subset (usually of size one) of components; a *conditional distribution* refers to events describing a subset of components when a second, disjoint subset is fixed to some conditioning event.

random variable: A random variable is a variable whose value follows a probability distribution. Random variables can take very different forms, such as discrete and continuous, or univariate and multivariate.

root node: A root node is a node in a DAG without any parent.

saturated graphs: A graph in which all nodes are adjacent to each others.

serial connection: A serial connection is one of the three fundamental connections between three nodes (see Figure 1.3 on page 21). Grandparent, parent, child: $A \rightarrow B \rightarrow C$.

skeleton: The skeleton of a (partially) directed graph is its underlying undirected graph, *i.e.*, the undirected graph obtained by disregarding all arc directions.

structure: The structure of a BN is the set of its conditional independencies; they are represented by its DAG.

topological order: An ordering on the nodes of a DAG is said to be topological when it is consistent with every directed path of the DAG. Often a topological order is not unique, and a DAG may admit more than one ordering.

tree: A tree is an undirected graph in which every pair of nodes is connected with exactly *one* path, or a directed path in which all nodes have exactly one parent except a single *root node*.

undirected graph: A graph comprising only undirected links.

v-structure: A v-structure is a convergent connection in which the two parents are not linked. Note that several v-structures can be centred on the same node (*i.e.*, have the same child node).

vertex/vertices: Synonymous for node(s).

Solutions

Exercises of Chapter 1

1.1 Consider the DAG for the survey studied in this chapter and shown in Figure 1.1.

1. List the parents and the children of each node.
2. List all the fundamental connections present in the DAG and classify them as either serial, divergent or convergent.
3. Add an arc from Age to Occupation, and another arc from Travel to Education. Is the resulting graph still a valid BN? If not, why?

1. Parents and children of each node are as follows.

   - Age (A) has no parent, and Education (E) is its only child.
   - Sex (S) has no parent, and Education (E) is its only child.
   - Education (E) has two parents, Age (A) and Sex (S), and two children, Occupation (O) and Residence (R).
   - Occupation (O) has one parent, Education (E), and one child, Travel (T).
   - Residence (R) has one parent, Education (E), and one child, Travel (T).
   - Travel (T) has two parents, Occupation (O) and Residence (R), and no child.

2. The fundamental connections in the DAG are as follows.

   - A → E ← S (convergent).
   - A → E → O (serial).
   - A → E → R (serial).
   - S → E → O (serial).
   - S → E → R (serial).
   - O ← E → R (divergent).
   - E → O → T (serial).
   - E → R → T (serial).
   - O → T ← R (convergent).

3. Adding A → O does not introduce any cycles; the graph is still a DAG and a valid BN. On the other hand, adding T → E introduces the following cycles: T → E → R → T and T → E → O → T. Therefore, the resulting graph is not acyclic and cannot be part of a valid BN.

227

1.2 Consider the probability distribution from the survey in Section 1.3.

1. Compute the number of configurations of the parents of each node.
2. Compute the number of parameters of the local distributions.
3. Compute the number of parameters of the global distribution.
4. Add an arc from Education to Travel. Recompute the factorisation into local distributions shown in Equation (1.1). How does the number of parameters of each local distribution change?

1. The number of parents' configurations is 2×3 for E, 2 for R, 2 for O and 4 for T. A and S have no parent.
2. The numbers of free parameters is 2 for A, 1 for S, 6×1 for E, 2×1 for O, 2×1 for R and 4×2 for T. Indeed, 1 must be removed from the number of probabilities since their sum is known to be one.
3. The number of free parameters of the global distribution is the sum of the numbers of free parameters of the conditional probability tables, that is, 21 parameters.
4. After adding E → T, Equation (1.1) reads

$$\Pr(A, S, E, O, R, T) =$$
$$\Pr(A)\Pr(S)\Pr(E \mid A, S)\Pr(O \mid E)\Pr(R \mid E)\Pr(T \mid E, O, R).$$

The number of free parameters of the local distribution of T increases to 2×8 due to the additional parents' configurations; all the other local distributions are unchanged.

1.3 Consider again the DAG for the survey.

1. Create an object of class bn for the DAG.
2. Use the functions in bnlearn and the R object created in the previous point to extract the nodes and the arcs of the DAG. Also extract the parents and the children of each node.
3. Print the model formula from bn.
4. Fit the parameters of the network from the data stored in survey.txt using their Bayesian estimators and save the result into an object of class bn.fit.
5. Remove the arc from Education to Occupation.
6. Fit the parameters of the modified network. Which local distributions change, and how?

1. The easiest way is to use Equation (1.1) and model2network as follows.

```
dag <- model2network("[A][S][E|A:S][O|E][R|E][T|O:R]")
```

2. ```
nodes(dag)
[1] "A" "E" "O" "R" "S" "T"
```
```

```
arcs(dag)
      from to
 [1,] "A"  "E"
 [2,] "S"  "E"
 [3,] "E"  "O"
 [4,] "E"  "R"
 [5,] "O"  "T"
 [6,] "R"  "T"
par <- sapply(nodes(dag), parents, x = dag)
chld <- sapply(nodes(dag), children, x = dag)
```

3.
```
modelstring(dag)
 [1] "[A][S][E|A:S][O|E][R|E][T|O:R]"
```

4.
```
survey <- read.table("../discrete/survey.txt", header = TRUE,
            colClasses = "factor")
fitted <- bn.fit(dag, survey, method = "bayes")
```

5.
```
dag2 <- drop.arc(dag, from = "E", to = "O")
```

6. The conditional probability table of O now has only two cells and one dimension, because O has no parent in dag2.

```
fitted2 <- bn.fit(dag2, survey, method = "bayes")
dim(coef(fitted$O))
 [1] 2 2
dim(coef(fitted2$O))
 [1] 2
```

1.4 Re-create the bn.mle object used in Section 1.4.

1. Compare the distribution of Occupation conditional on Age with the corresponding marginal distribution using querygrain.
2. How many random observations are needed for cpquery to produce estimates of the parameters of these two distributions with a precision of ±0.01?
3. Use the functions in bnlearn to extract the DAG from bn.mle.
4. Which nodes d-separate Age and Occupation?

1. Conditioning does not have a big effect on the distribution of O.

```
library(gRain)
junction <- compile(as.grain(bn.mle))
querygrain(junction, nodes = "O")$O
O
   emp  self
 0.966 0.034
jage <- setEvidence(junction, "A", states = "young")
```

```
querygrain(jage, nodes = "O")$O
  O
    emp   self
  0.9644 0.0356
jage <- setEvidence(junction, "A", states = "adult")
querygrain(jage, nodes = "O")$O
  O
    emp   self
  0.9636 0.0364
jage <- setEvidence(junction, "A", states = "old")
querygrain(jage, nodes = "O")$O
  O
    emp   self
  0.9739 0.0261
```

2. 10^3 simulations are enough for likelihood weighting, 10^4 for logic sampling.

```
set.seed(123)
cpquery(bn.mle, event = (O == "emp"),
  evidence = list(A = "young"), method = "lw",
  n = 10^3)
  [1] 0.96
cpquery(bn.mle, event = (O == "emp"),
  evidence = (A == "young"), method = "ls",
  n = 10^4)
  [1] 0.969
```

3. `dag <- bn.net(bn.mle)`

4. `sapply(nodes(dag), function(z) dsep(dag, "A", "O", z))`

```
     A    E    O     R     S     T
  TRUE TRUE TRUE FALSE FALSE FALSE
```

1.5 Implement an R function for BN inference via rejection sampling using the description provided in Section 1.4 as a reference.

```
rejection.sampling <- function(bn, nsim, event.node,
    event.value, evidence.node, evidence.value) {

  sims <- rbn(bn, nsim)
  m1 <- sims[sims[, evidence.node] == evidence.value, ]
  m2 <- m1[m1[, event.node] == event.value, ]
  return(nrow(m2)/nrow(m1))

}#REJECTION.SAMPLING
rejection.sampling(bn.mle, nsim = 10^4, event.node = "O",
  event.value = "emp", evidence.node = "A",
  evidence.value = "young")
  [1] 0.96
```

1.6 Using the dag and bn objects from Sections 1.2 and 1.3:

1. Plot the DAG using graphviz.plot.
2. Plot the DAG again, highlighting the nodes and the arcs that are part of one or more v-structures.
3. Plot the DAG one more time, highlighting the path leading from Age to Occupation.
4. Plot the conditional probability table of Education.
5. Compare graphically the distributions of Education for male and female interviewees.

1. ```
   graphviz.plot(dag)
   ```

2. ```
   vs <- vstructs(dag, arcs = TRUE)
   hl <- list(nodes = unique(as.character(vs)), arcs = vs)
   graphviz.plot(dag, highlight = hl)
   ```

3. ```
   hl <- matrix(c("A", "E", "E", "O"), nc = 2,
           byrow = TRUE)
   graphviz.plot(dag, highlight = list(arcs = hl))
   ```

4. ```
   bn.fit.barchart(bn$E)
   ```

5. ```
   library(gRain)
   junction <- compile(as.grain(bn))
   jmale <- setEvidence(junction, "S", states = "M")
   jfemale <- setEvidence(junction, "S", states = "F")
   library(lattice)
   library(gridExtra)
   p1 <- barchart(querygrain(jmale, nodes = "E")$E,
           main = "Male", xlim = c(0, 1))
   p2 <- barchart(querygrain(jfemale, nodes = "E")$E,
           main = "Female", xlim = c(0, 1))
   grid.arrange(p1, p2, ncol = 2)
   ```

Exercises of Chapter 2

2.1 Prove that Equation (2.2) implies Equation (2.3).

Using Bayes' theorem twice, we obtain that

$$
\begin{aligned}
f\left(G = g \mid C = c\right) &= \frac{f\left(G = g, C = c\right)}{f\left(C = c\right)} \\
&= \frac{f\left(C = c \mid G = g\right)}{f\left(C = c\right)} f\left(G = g\right).
\end{aligned}
$$

If $f\left(G = g \mid C = c\right) = f\left(G = g\right)$ it would imply that $f\left(C = c \mid G = g\right) = f\left(C = c\right)$ which contradicts the first step in the proof.
   ```

2.2   Within the context of the DAG shown in Figure 2.1, prove that Equation (2.5) is true using Equation (2.6).

From Equation (2.6), we can obtain f (V, W, N) integrating out the other variables:

$$
\begin{aligned}
f(V, W, N) &= \int_G \int_E \int_C f(G, E, V, N, W, C) \\
&= f(V) f(N \mid V) f(W \mid V) \times \\
&\quad \left(\int_G \int_E f(G) f(E) f(V \mid G, E) \right) \left(\int_C f(C \mid N, W) \right) \\
&= f(V) f(N \mid V) f(W \mid V)
\end{aligned}
$$

and from f (V, W, N) we can obtain the joint distribution of W and N conditional to V

$$
\begin{aligned}
f(W, N \mid V) &= \frac{f(V, W, N)}{f(V)} \\
&= f(N \mid V) f(W \mid V)
\end{aligned}
$$

characteristic of the conditional independence.

2.3   Compute the marginal variance of the two nodes with two parents from the local distributions proposed in Table 2.1. Why is it much more complicated for C than for V?

V and C are the two nodes for which we have to compute the variance. For the sake of clarity, we will first replace the numeric values of the coefficient with arbitrary constants k_0, k_1, k_2 and k_3. From the first three equations of Table 2.1, V can be written as

$$
V = k_0 + k_1 G + k_2 E + k_3 \varepsilon_V,
$$

where G, E and ε_V are independent random variables with variances 10^2, 10^2 and 5^2, respectively, so we can easily find that

$$
\begin{aligned}
\mathrm{VAR}(V) &= k_1^2 \, \mathrm{VAR}(G) + k_2^2 \, \mathrm{VAR}(E) + k_3^2 \, \mathrm{VAR}(\varepsilon_V) \\
&= \left(\frac{1}{2}\right)^2 10^2 + \left(\sqrt{\frac{1}{2}}\right)^2 10^2 + 5^2 \\
&= 10.
\end{aligned}
$$

For C, things are a little more complicated because

$$
C = k_1 N + k_2 W + k_3 \varepsilon_C
$$

and the variables N and W are not independent, as they share V as a common ancestor. Therefore we have to first compute COV (N, V):

$$
\begin{aligned}
\mathrm{COV}(N, V) &= \mathrm{COV}(h_1 V, h_2 V) \\
&= h_1 h_2 \, \mathrm{VAR}(V) \\
&= 0.1 \times 0.7 \times 10^2
\end{aligned}
$$

and finally

$$
\begin{aligned}
\mathrm{VAR}(C) &= k_1^2 \, \mathrm{VAR}(N) + k_2^2 \, \mathrm{VAR}(W) + k_3^2 \, \mathrm{VAR}(\varepsilon_C) + 2 k_1 k_2 \, \mathrm{COV}(N, W) \\
&= \left(0.3^2 + 0.7^2 + 0.3 \times 0.7 \times 0.14\right) 10^2 + 6.25^2 \\
&= (10.00012)^2.
\end{aligned}
$$

2.4 Write an R script using only the `rnorm` **and** `cbind` **functions to create a** 100×6 **matrix of** 100 **observations simulated from the BN defined in Table 2.1. Compare the result with those produced by a call to** `cpdist` **function.**

First, it is sensible to initialise the pseudo-random generator and parameterise the desired number of simulations.

```
set.seed(12345)
ns <- 100
```

Then the following seven commands answer the first question.

```
sG <- rnorm(ns, 50, 10)
sE <- rnorm(ns, 50, 10)
sV <- rnorm(ns, -10.35534 + 0.5 * sG + 0.70711 * sE, 5)
sN <- rnorm(ns, 45 + 0.1 * sV, 9.949874)
sW <- rnorm(ns, 15 + 0.7 * sV, 7.141428)
sC <- rnorm(ns, 0.3 * sN + 0.7 * sW, 6.25)
simu1 <- cbind(sG, sE, sV, sN, sW, sC)
```

To perform the equivalent simulation with `cpdist`, we first have to define the BN:

```
library(bnlearn)
dag.bnlearn <- model2network("[G][E][V|G:E][N|V][W|V][C|N:W]")
disE <- list(coef = c("(Intercept)" = 50), sd = 10)
disG <- list(coef = c("(Intercept)" = 50), sd = 10)
disV <- list(coef = c("(Intercept)" = -10.35534,
             E = 0.70711, G = 0.5), sd = 5)
disN <- list(coef = c("(Intercept)" = 45, V = 0.1), sd = 9.949874)
disW <- list(coef = c("(Intercept)" = 15, V = 0.7), sd = 7.141428)
disC <- list(coef = c("(Intercept)" = 0, N = 0.3, W = 0.7), sd = 6.25)
dis.list <- list(E = disE, G = disG, V = disV, N = disN,
                 W = disW, C = disC)
gbn.bnlearn <- custom.fit(dag.bnlearn, dist = dis.list)
simu2 <- cpdist(gbn.bnlearn, nodes = nodes(gbn.bnlearn),
          evidence = TRUE)
```

The comparison can be made in different ways; here we will just extract the medians for all nodes.

```
su1 <- summary(simu1)
su2 <- summary(simu2)
cbind(simu1 = su1[3, ], simu2 = su2[3, ])
      simu1             simu2
  sG "Median :54.8  " "Median :50.0  "
  sE "Median :50.3  " "Median :49.9  "
  sV "Median :49.9  " "Median :49.9  "
  sN "Median :53.6  " "Median :50.2  "
  sW "Median :49.3  " "Median :49.9  "
  sC "Median :50.7  " "Median :50.0  "
```

Indeed even if all results are around the expected value of 50, there are differences; but they are comparable within each column and between different columns. They are the consequence of the small size of the simulation, making the same calculation with $\mathtt{ns} = 10^4$ gives a more precise result, even if still approximate.

2.5 Imagine two ways other than changing the size of the points (as in Section 2.7.2) to introduce a third variable in the plot.

This is not an easy question. Many approaches have been proposed and the user has to choose that best suited to the data. Here, denoting with $(\mathtt{A}, \mathtt{B}, \mathtt{C})$ the three variables we want to explore, we illustrate three possible choices.

- The simplest choice is perhaps to make three scatter plots of two variables each, dividing the plot into four quarters.

  1. Making plot(A, B) into the top right quarter,
  2. making plot(A, C) into the bottom right quarter (A on the same scale as in the previous plot),
  3. making plot(B, C) into the bottom left quarter (C on the same scale as in the previous plot).

  The top-left quarter is left empty.

- Another choice is to associate a different symbol or shape to each point depending on the values of the three variables. A possible example is the figure shown in Hartigan (1975, page 39), in which each point is associated with a rectangular box having the values of the three variables (assumed to be positive) as dimensions. Some points will be represented with tall boxes, others with flat boxes; the depth is also important.

- Finally, we can create a true 3D plot and rotate it interactively to assess the distance between the points. Suitable functions are available in R, for instance in the **rgl** package.

2.6 Can GBNs be extended to log-normal distributions? If so how, if not, why?

Of course this is possible and very easy! Just take the logarithm of the initial variables and apply a GBN to the transformed variables. There is no need to do that for all variables; we can transform only some of them. Note that it is necessary that all possible values of the variables to transform are positive. From a practical point of view, a constant can be added to satisfy this constraint, giving access to a third parameter for

$$A \to \log\left(A + k\right) \sim N\left(\mu, \sigma^2\right). \tag{C.1}$$

In general any univariate or multivariate transformation can be applied to a set of variables before attempting a model based on a GBN.

2.7 How can we generalise GBNs as defined in Section 2.3 in order to make each node's variance depend on the node's parents?

Indeed a strong restriction of GBNs is that the conditional variance of a node with respect to its parents must be a constant. There is no difficulty in proposing a generalisation. For instance, if A is the unique parent of B, we could assume that

$$\mathrm{VAR}(\mathtt{B} \mid \mathtt{A}) = (\mathtt{A} - \mathrm{E}(\mathtt{A}))^2 \times \sigma_B^2.$$

This is not usually done because most of BN's methodological developments are very general and do not discuss specific, limited cases of ad-hoc node parameterisations such as most instances of hybrid BNs.

2.8 From the first three lines of Table 2.1, prove that the joint distribution of E, G and V is trivariate normal.

To prove it, we will use the following result: the logarithm of the density of a multivariate normal distribution is, up to an additive constant:

$$f(\mathbf{x}) \propto -\frac{1}{2}(\mathbf{x} - \boldsymbol{\mu})^T \Sigma^{-2} (\mathbf{x} - \boldsymbol{\mu})$$

where $\mathbf{x}$ stands for the value of the random vector, $\boldsymbol{\mu}$ is its expectation and Σ its covariance matrix.

To simplify the notation, we will first transform the three variables to give them a zero marginal expectation and a unity marginal variance.

$$\widetilde{G} = \frac{G - E(G)}{\sqrt{\text{VAR}(G)}} = \frac{G - 50}{10},$$

$$\widetilde{E} = \frac{E - E(E)}{\sqrt{\text{VAR}(E)}} = \frac{E - 50}{10},$$

$$\widetilde{V} = \frac{V - E(V)}{\sqrt{\text{VAR}(V)}} = \frac{V - 50}{10}.$$

The resulting normalised variables are

$$\widetilde{G} \sim N(0,1),$$
$$\widetilde{E} \sim N(0,1),$$
$$\widetilde{V} \mid \widetilde{G}, \widetilde{E} \sim N\left(\frac{1}{2}\widetilde{G} + \sqrt{\frac{1}{2}}\widetilde{E}, \left(\frac{1}{2}\right)^2\right).$$

We are now able to compute the joint density of the three transformed variables

$$f\left(\widetilde{G} = g, \widetilde{E} = e, \widetilde{V} = v\right) \quad \propto \quad f\left(\widetilde{G} = g\right) + f\left(\widetilde{E} = e\right) + f\left(\widetilde{V} = v \mid \widetilde{G} = g, \widetilde{E} = e\right)$$

$$= \quad -\frac{g^2}{2} - \frac{e^2}{2} - 2\left(v - \frac{1}{2}g - \sqrt{\frac{1}{2}}e\right)^2$$

$$= \quad -\begin{bmatrix} g \\ e \\ v \end{bmatrix}^T \begin{bmatrix} 1 & \frac{\sqrt{2}}{2} & -1 \\ \frac{\sqrt{2}}{2} & \frac{3}{2} & -\sqrt{2} \\ -1 & -\sqrt{2} & 2 \end{bmatrix} \begin{bmatrix} g \\ e \\ v \end{bmatrix}$$

$$= \quad -\frac{1}{2}\begin{bmatrix} g \\ e \\ v \end{bmatrix}^T \begin{bmatrix} 1 & 0 & \frac{1}{2} \\ 0 & 1 & \sqrt{\frac{1}{2}} \\ \frac{1}{2} & \sqrt{\frac{1}{2}} & 1 \end{bmatrix}^{-1} \begin{bmatrix} g \\ e \\ v \end{bmatrix}.$$

Now

$$\text{VAR}\left(\begin{bmatrix} \widetilde{G} \\ \widetilde{E} \\ \widetilde{V} \end{bmatrix}\right) = \begin{bmatrix} 1 & 0 & \frac{1}{2} \\ 0 & 1 & \sqrt{\frac{1}{2}} \\ \frac{1}{2} & \sqrt{\frac{1}{2}} & 1 \end{bmatrix}$$

which completes the proof.

Exercises of Chapter 3

3.1 Consider the expression for BIC in Equation (3.4). What are the values of d_A, d_C, d_H, d_D, d_I, d_O and d_T? Or, equivalently, how many parameters does each local distribution have? How many (fewer) parameters would the corresponding CGBN fitted with complete pooling have?

$d_A = 2$, $d_C = 6$, $d_H = 3$, $d_D = 12$, $d_I = 9$, $d_O = 6$ and $d_T = 4$, for a total of $d = 42$ parameters.

Complete pooling means that I, O and D are fitted with a single regression model each, as opposed to a mixture of regressions each. After expanding the discrete variables into contrasts, $d_I = 7$, $d_O = 4$ and $d_O = 7$. Hence we would save 9 parameters out of the 42 of the original model.

3.2 The deal package contains a data set called ksl, which describes the health and social characteristics of a sample of Danish 70 year olds. It includes the following variables: FEV (*Forced ejection volume*), Kol (*Cholesterol*), Hyp (*Hypertension*), logBMI (*Logarithm of Body Mass Index*), Smok (*Smoking*), Alc (*Alcohol consumption*), Work (*Working*), Sex an Year.

1. Load the ksl data, ensure that all variables are either "factor" or "numeric", and use hc to learn the structure of a BN.

2. Learn the parameters of the BN.

3. Use cpquery to compute the probability of having hypertension in the general population, in individuals with cholesterol higher than 750, and in individuals with log-BMI higher than 3.5. Are these probabilities comparable?

4. Use cpdist to simulate the joint distribution of cholesterol and log-BMI for individuals that smoke but have no hypertension. Are these two variable correlated?

1.
```r
library(deal)
data(ksl)
ksl$FEV <- as.numeric(ksl$FEV)
ksl$Kol <- as.numeric(ksl$Kol)
ksl$Hyp <- factor(ksl$Hyp, labels = c("no", "yes"))
ksl$Smok <- factor(ksl$Smok, labels = c("no", "yes"))
dag <- hc(ksl, score = "bic-cg")
```

2.
```r
fitted <- bn.fit(dag, data = ksl)
```

3.
```r
cpquery(fitted, event = (Hyp == "yes"), evidence = TRUE)
[1] 0.554
cpquery(fitted, event = (Hyp == "yes"), evidence = (Kol > 750))
[1] 0.576
cpquery(fitted, event = (Hyp == "yes"), evidence = (logBMI > 3.5))
[1] 0.881
```

Individuals with high cholesterol have a comparable probability of having hypertension to the general population. Individuals with high log-BMI has a markedly higher probability of having hypertension.

4.
```
part <- cpdist(fitted, nodes = c("Kol", "logBMI"),
               evidence = (Smok == "yes") & (Hyp == "no"))
cor(part$Kol, part$logBMI)
  [1] -0.015
```

Kol and logBMI do not appear to be significantly correlated.

3.3 One of the first examples of mixed-effects models in Pinheiro and Bates (2000) describes a randomised-block experiment which recorded the effort (E) required by each of nine different subjects (S) to arise from each of four types of stools (T).

1. **Load the ergoStool data for this experiment from the nlme package, and rename the variables to E, S, T.**
2. **Fit a mixed-effects model in which T is a fixed effect, S is a random intercept effect and E is the response variable; and the corresponding classic linear model in which T and S are the explanatory variables.**
3. **Learn the parameters of the CGBN with DAG S→E←T from ergoStool.**
4. **Comment on the differences and the similarities in the parameterisations of these three models.**

1.
```
library(nlme)
names(ergoStool) <- c("E", "T", "S")
```

2.
```
lm(E ~ T + S, data = ergoStool)
lme(E ~ T, random = ~ 1 | S, data = ergoStool)
```

3.
```
dag <- model2network("[E|T:S][T][S]")
fitted <- bn.fit(dag, ergoStool)
```

4. All these models assume that residuals are normally distributed, but they vary in what variance they assume these residuals have. The CGBN assumes a different variance for each T and S; the mixed effects model will give a different variance for each S (the random effect); the classic linear models assumes all residuals have the same variance.

They also differ in the parameterisation of the regression coefficients. Both the mixed-effects model and the classic linear regression expand T into contrasts and incorporate the first of them into the intercept; while the CGBN models all levels of T explicitly. The random-effects model also forces the regression coefficients associated with the random intercept S to sum up to zero, which is not the case for the classic linear regression or for the CGBN.

Exercises of Chapter 4

4.1 Consider the networks in Figure 4.1.

1. How many parameters have the networks, and the local distributions associated with the individual nodes, in the top-left, top-right and bottom-left panels?

2. Extend the network in the bottom-left panel to model as second time point t_2 in addition to t_0 and t_1. How many additional parameters does that require?

3. Finally, make the nodes in t_2 dependent on the nodes in t_0 in addition to those in t_1. How many additional parameters does that require?

1. The network in the top left-panel has

$$\underbrace{(2-1)}_{\text{W}} + \underbrace{(3-1)}_{\text{Tout}} + \underbrace{2 \times (2-1)}_{\text{St}} + \underbrace{3 \times 2 \times (3-1)}_{\text{Tin}} = 17 \text{ parameters.}$$

The network in the top-right panel has

$$17 + \underbrace{2 \times (2-1)}_{\text{W1}} + \underbrace{3 \times (3-1)}_{\text{Tout1}} + \underbrace{2 \times 2 \times (2-1)}_{\text{St1}} +$$

$$\underbrace{3 \times 2 \times 3 \times (3-1)}_{\text{Tin1}} = 17 + 48 = 65 \text{ parameters.}$$

The network in the bottom-left panel has

$$\underbrace{(2-1)}_{\text{W0}} + \underbrace{(3-1)}_{\text{Tout0}} + \underbrace{(3-1)}_{\text{Tin0}} + \underbrace{(2-1)}_{\text{St0}} + \underbrace{2 \times 2 \times (2-1)}_{\text{St1}} + \underbrace{3 \times (3-1)}_{\text{Tout1}} +$$

$$\underbrace{3 \times 3 \times 2 \times (3-1)}_{\text{Tin1}} = 6 + 46 = 52 \text{ parameters.}$$

As mentioned in the text, this last network has fewer parameters than the previous one since it does not include the arcs between the variables in t_0.

2. Including the additional nodes St2, Tout2, Tin2 to model t_2 conditional on t_1 adds 46 more parameters to the DBN, since we need to model the transition from t_1 to t_2 without assuming it is identical to that from t_0 to t_1.

3. Making St2, Tout2, Tin2 depend respectively on StZ, ToutZ, TinZ in addition to St, Tout, Tin adds

$$\underbrace{2 \times 2 \times 2 \times (2-1)}_{\text{St2}} + \underbrace{3 \times 3 \times (3-1)}_{\text{Tout2}} + \underbrace{3 \times 3 \times 3 \times 2 \times (3-1)}_{\text{Tin2}} = 134$$

parameters. The increase in much larger than in point 2.

4.2 Consider again the DBN in the bottom-left panel of Figure 4.1.

1. Extend the network to model t_2 as in point 2 of Exercise 4.1, and create the bn object encoding it. Call the new nodes St2, Tin2 and Tout2.
2. Use the conditional probabilities from Section 4.3 to create a bn.fit object from the bn object in the previous point.
3. Use cpquery to compute the probability that Tin2 is equal to "18-24" and St2 is "low" given that Tin0 is "18-24" and St0 is "high" when windows are either open or closed. What would you expect compared to the similar query we performed in Section 4.5?

```
1.  dag <- model2network(paste0("[St0][W0][Tout0][Tin0][St1|St0:W0]",
            "[Tout1|Tout0][Tin1|Tin0:Tout1:W0][St2|St1:W0][Tout2|Tout1]",
            "[Tin2|Tin1:Tout2:W0]"))
```

```
2.  St2.prob <- St1.prob
    names(dimnames(St2.prob)) <- c("St2", "St1", "W0")
    Tout2.prob <- Tout1.prob
    names(dimnames(Tout2.prob)) <- c("Tout2", "Tout1")
    Tin2.prob <- Tin1.prob
    names(dimnames(Tin2.prob)) <- c("Tin2", "Tout2", "W0", "Tin1")
    cpts <- list(St0 = St0.prob, W0 = W0.prob, Tout0 = Tout0.prob,
            Tin0 = Tin0.prob, St1 = St1.prob, Tout1 = Tout1.prob,
            Tin1 = Tin1.prob, St2 = St2.prob, Tout2 = Tout2.prob,
            Tin2 = Tin2.prob)
    fitted <- custom.fit(dag, cpts)
```

```
3.  cpquery(fitted, event = (St2 == "low") & (Tin2 == "18-24"),
      evidence = (St0 == "high") & (Tin0 == "18-24") & (W0 == "closed"))
    [1] 0.0153
    cpquery(fitted, event = (St2 == "low") & (Tin2 == "18-24"),
      evidence = (St0 == "high") & (Tin0 == "18-24") & (W0 == "open"))
    [1] 0.483
```

As expected the probability that air quality gets better is higher, since the windows have been left open for longer. If windows are kept closed then there is no variation in air quality.

4.3 Consider the egl.DSE.data in the dse package, and in particular the three output series M1, GDP and CPI indexes for Canada between 1961 and 1991.

1. Load the data and reformat them to have variables MI_0, GDPl2_0, CPI_0, MI_1, GDPl2_1, CPI_1.
2. Learn the structure of DBN using the appropriate blacklists.
3. Learn the parameters of the DBN modelling it as GBN. What kind of model do you get as a result?
4. Revisit the first two points to learn the structure of a DBN spanning t_0, t_1 and t_2.

1. ```
   library(dse)
   data(eg1.DSE.data)
   mts <- as.data.frame(eg1.DSE.data$output)
   mts <- data.frame(mts[-nrow(mts), ], mts[-1, ])
   t0.nodes <- c("M1_0", "GDPl2_0", "CPI_0")
   t1.nodes <- c("M1_1", "GDPl2_1", "CPI_1")
   names(mts) <- c(t0.nodes, t1.nodes)
   ```

2. ```
   bl <- tiers2blacklist(list(t0.nodes, t1.nodes))
   bl <- rbind(bl, set2blacklist(t0.nodes))
   dag <- hc(mts, blacklist = bl)
   ```

3. The local distributions are linear regression of the target nodes against
   its parents, giving a model that is very similar to a *vector auto-regressive*
   model

   ```
   fitted <- bn.fit(dag, mts)
   ```

4. ```
   mts <- as.data.frame(eg1.DSE.data$output)
   mts <- data.frame(mts[-c(nrow(mts), nrow(mts) - 1), ],
           mts[-c(1, nrow(mts)), ], mts[-(1:2), ])
   t0.nodes <- c("M1_0", "GDPl2_0", "CPI_0")
   t1.nodes <- c("M1_1", "GDPl2_1", "CPI_1")
   t2.nodes <- c("M1_2", "GDPl2_2", "CPI_2")
   names(mts) <- c(t0.nodes, t1.nodes, t2.nodes)
   bl <- tiers2blacklist(list(t0.nodes, t1.nodes, t2.nodes))
   bl <- rbind(bl, set2blacklist(t0.nodes))
   dag <- hc(mts, blacklist = bl)
   ```

Exercises of Chapter 5

5.1 Consider again the ergoStool data from Exercise 3.3, which is originally
from Pinheiro and Bates (2000) and which describes a randomised-block
experiment which recorded the effort (E) required by each of nine different
subjects (S) to arise from each of four types of stools (T).

1. Load the ergoStool data for this experiment from the nlme
   package, and rename the variables to E, S, T.
2. Create a BN in Stan in which S, T are multinomials and E is a
   normal random variable, and use the ergoStool data to estimate
   their parameters and sample from their posterior distributions.
   The structure of the BN should be $S \rightarrow E \leftarrow T$.
3. Summarise the posterior samples with their mean and standard
   deviation, and comment on the precision of the posterior
   estimates.
   ```

1. ```
   library(nlme)
   names(ergoStool) <- c("E", "T", "S")
   ```

2. ```
   TSm <- model.matrix(~ T + S, data = ergoStool)
   S <- as.integer(ergoStool$S)
   T <- as.integer(ergoStool$T)

   stancode <- 'data {
     int<lower=0> n;
     vector[n] E;
     int S[n];
     int T[n];
     matrix[n, 12] TSm;
   }
   parameters {
     simplex[4] Tp;
     simplex[9] Sp;
     vector[12] beta;
     real<lower=0> sigma;
   }
   model {
     T ~ categorical(Tp);
     S ~ categorical(Sp);
     E ~ normal(TSm * beta, sigma);
   }'

   parameters.model <- stan_model(model_code = stancode)
   fit <- sampling(parameters.model, iter = 5000, thin = 25,
             data = list(n = nrow(ergoStool), E = ergoStool$E,
                 T = T, S = S, TSm = TSm))
   ```

3. Comparing the magnitude of the standard deviations with that of the corresponding means suggests that a number of parameter estimates are not very accurate and have wide credible intervals. This can be attributed to a combination of small sample size and non-informative priors.

   ```
   fit <- as.data.frame(fit)
   fit <- fit[, names(fit) != "lp__"]
   est <- data.frame(mean = sapply(fit, mean), sd = sapply(fit, sd))
   ```

5.2 Consider a bus stop with two bus lines B1 and B2, and a tram line T. Buses on these lines have a number of passengers that can be modelled as $Pois(\lambda_1)$ and $Pois(\lambda_2)$, respectively; passengers may decide to get off with probabilities p_1 and p_2. Trams will have $Pois(\lambda_3)$ passengers, which may decide to get off with probability p_3. Passengers disembarked from B1 and B2 may then decide to board T with probabilities q_1 and q_2.

Create a Stan model that can be used to simulate how many passengers will be on the tram when it departs, assuming $\lambda_1 = \lambda_2 = 25$, $\lambda_3 = 20$, $p_1 = 0.6$, $p_2 = 0.3$, $p_3 = 0.1$, $q_1 = q_2 = 0.15$.

```
stancode <- 'data{
  real lambda[3];
  real p[3];
  real q[2];
}
generated quantities{
  int B1;
  int B2;
  int T;
  int dis1;
  int dis2;
  int disT;
  int out;
  B1 = poisson_rng(lambda[1]);
  B2 = poisson_rng(lambda[2]);
  dis1 = binomial_rng(B1, p[1]);
  dis2 = binomial_rng(B2, p[2]);
  T = poisson_rng(lambda[3]);
  disT = binomial_rng(T, p[3]);
  out = T - disT + binomial_rng(dis1, q[1]) + binomial_rng(dis2, q[2]);
}'
model <- stan_model(model_code = stancode)
params <- list(lambda = c(25, 25, 20), p = c(0.6, 0.3, 0.1),
            q = c(0.15, 0.15))
fit <- sampling(model, algorithm = "Fixed_param",
        data = params, thin = 25, iter = 5000, seed = 42)
```

The distribution of the passengers on the tram is as follows.

```
df <- as.data.frame(extract(fit))
summary(df$T)
   Min. 1st Qu.  Median    Mean 3rd Qu.    Max.
    9.0    17.0    20.0    20.1    23.0    36.0
```

5.3 Consider again the bus and tram lines from Exercise 5.2. Construct a Stan model that simulates from the posterior distribution of the parameters assuming normal priors for λ_1, λ_2, λ_3 and beta priors for p_1, p_2, p_3, q_1 and q_2. Use the data simulated in Exercise 5.2 to run the model.

```
stancode <- 'data {
  int<lower=0> n;
  int<lower=0> B1[n];
  int<lower=0> B2[n];
  int<lower=0> dis1[n];
  int<lower=0> dis2[n];
  int<lower=0> dis3[n];
  int<lower=0> T[n];
  int<lower=0> out1[n];
  int<lower=0> out2[n];
}
```

```
parameters {
  real<lower=0> lambda[3];
  vector<lower=0,upper=1>[3] p;
  vector<lower=0,upper=1>[2] q;
}
model {
  for (i in 1:3)
    lambda[i] ~ normal(25, 2);
  for (i in 1:3)
    p[i] ~ beta(2, 2);
  for (i in 1:2)
    q[i] ~ beta(2, 5);
  B1 ~ poisson(lambda[1]);
  B2 ~ poisson(lambda[2]);
  dis1 ~ binomial(B1, p[1]);
  dis2 ~ binomial(B2, p[2]);
  T ~ poisson(lambda[3]);
  dis3 ~ binomial(T, p[3]);
  out1 ~ binomial(dis1, q[1]);
  out2 ~ binomial(dis2, q[2]);
}'
```

Exercises of Chapter 6

6.1 Consider the network below.

```
dag <- model2network("[A][C|A:B][D|C:A][F|A][B][E|B:D:F]")
```

1. What are the minimal set(s) of variables required to d-separate C and E (that is, sets of variables for which no proper subset d-separates C and E)?

2. What are the minimal set(s) of variables required to d-separate A and B?

3. What are the maximal set(s) of variables that d-separate C and E (that is, sets of variables for which no proper superset d-separates C and E)?

4. What are the maximal set(s) of variables that d-separate A and B?

5. Using the chain rule, establish an expression for the joint distribution over {A, B, C, D, E, F}. Use this expression to show that B and D are conditionally independent given A and C.

Minimal sets d-separating C and E: $\{B, D, F\}$ and $\{A, B, D\}$.

```
nodes <- nodes(dag)

for (size in 1:(length(nodes) - 2)) {

  candidates <- setdiff(nodes, c("C", "E"))
  sets <- combn(candidates, size, simplify = FALSE)

  for (i in 1:length(sets))
    if (dsep(dag, "C", "E", sets[[i]]))
      cat("C and E are d-separated by", sets[[i]], "\n")

}#FOR
```
```
 C and E are d-separated by A B D
 C and E are d-separated by B D F
 C and E are d-separated by A B D F
```

Minimal sets d-separating A and B: $\emptyset$.
Maximal set d-separating C and E: $\{A, B, D, F\}$.
Maximal set d-separating A and B: $\{F\}$.
The network gives the local distributions:

$$\Pr(A, B, C, D, E, F) = \Pr(A)\Pr(B)\Pr(C \mid A, B)\Pr(D \mid A, C)\Pr(E \mid B, D, F)\Pr(F \mid A).$$

For proving that B and D are conditionally independent given A and C:

$$\Pr(B, D, A = a, C = c)$$
$$= \sum_E \sum_F \Pr(A = a)\Pr(B)\Pr(C = c \mid A = a, B)\Pr(D \mid A = a, C = c) \cdot$$
$$\Pr(E \mid B, D, F)\Pr(F \mid A = a)$$
$$= \Pr(A = a)\Pr(B)\Pr(C = c \mid A = a, B)\Pr(D \mid A = a, C = c) \cdot$$
$$\sum_E \sum_F \Pr(E \mid B, D, F)\Pr(F \mid A = a)$$
$$= \Pr(A = a)\Pr(B)\Pr(C = c \mid A = a, B)\Pr(D \mid A = a, C = c)$$

Then:

$$\Pr(B \mid D, A = a, C = c) = \frac{\Pr(A = a)\Pr(B)\Pr(C = c \mid A = a, B)\Pr(D \mid A = a, C = c)}{\sum_B \Pr(A = a)\Pr(B)\Pr(C = c \mid A = a, B)\Pr(D \mid A = a, C = c)}$$
$$= \frac{\cancel{\Pr(A = a)}\Pr(B)\Pr(C = c \mid A = a, B)\cancel{\Pr(D \mid A = a, C = c)}}{\cancel{\Pr(A = a)}\cancel{\Pr(D \mid A = a, C = c)}\sum_B \Pr(B)\Pr(C = c \mid A = a, B)}$$
$$= \frac{\Pr(B)\Pr(C = c \mid A = a, B)}{\sum_B \Pr(B)\Pr(C = c \mid A = a, B)}$$
$$= \frac{\Pr(B, C = c \mid A = a)}{\sum_B \Pr(B, C = c \mid A = a)}$$
$$= \Pr(B \mid A = a, C = c).$$

That is, we have $\Pr(\mathsf{B} \mid \mathsf{D}, \mathsf{A} = a, \mathsf{C} = c) = \Pr(\mathsf{B} \mid \mathsf{A} = a, \mathsf{C} = c)$, for an arbitrary choice of a and c, and thus that B and D are conditionally independent given A and C.

6.2 Let A be a node in a DAG. Assume that all variables in the Markov blanket of A are instantiated. Show that A is d-separated from the remaining uninstantiated variables.

Consider a path from A to B, and let C be the neighbour to A on that path. Then C is instantiated. If the C-connection is serial or diverging, then C blocks the path. If the link is converging, consider C's neighbour, D, on the path. As A and D have the common child C, D is in the Markov blanket for A and must be instantiated. Since the direction on the (D, C) link is from D to C, the D-connection on the path from A through C and D to B cannot be converging, and therefore D blocks the path.

6.3 Consider the following DAGs.

```
dag1 <- model2network("[G][H][D|G][I|G:H][F|H][B|D][E|G][C|E:F][A|B:C]")
dag2 <-
  model2network("[B][C][A][D|B][E|B][F|C][G|A:D][H|D:E][I|E:F][J|G]")
```

1. Determine which variables are d-separated from A given B (in dag1) and which variables are d-separated from A given J (in dag2).
2. Apply the procedure shown in the Section 6.4 to determine whether A is d-separated from F given $\{\mathsf{H}, \mathsf{I}\}$ in dag1.
3. Similarly, determine whether A is d-separated from C given B in dag2.

1.  ```
    sapply(nodes(dag1), dsep, bn = dag1, y = "A", z = "B")
        A     B     C     D     E     F     G     H     I
    FALSE  TRUE FALSE FALSE FALSE FALSE FALSE FALSE FALSE
    sapply(nodes(dag2), dsep, bn = dag2, y = "A", z = "J")
        A     B     C     D     E     F     G     H     I     J
    FALSE FALSE  TRUE FALSE FALSE  TRUE FALSE FALSE FALSE  TRUE
    ```

    Obviously, A cannot be d-separated from itself and B, J are always d-separated from A since they are part of the d-separating set.

2. There exists a path between A and F in the reduced moral graph (from which the putative d-separating set has been removed), so A and F are not d-separated by H and I.

    ```
    ancA <- ancestors(dag1, "A")
    ancF <- ancestors(dag1, "F")
    ancHI <- union(ancestors(dag1, "H"), ancestors(dag1, "I"))
    sub <- subgraph(dag1,
            unique(c("A", "F", "H", "I", ancA, ancF, ancHI)))
    mo <- moral(sub)
    msub <- subgraph(mo, setdiff(nodes(mo), c("H", "I")))
    path.exists(mo, from = "A", to = "F")
    [1] TRUE
    ```
    ```

3. The ancestral graph consists of three isolated nodes, A, B and C.

```
ancA <- ancestors(dag2, "A")
ancB <- ancestors(dag2, "B")
ancC <- ancestors(dag2, "C")
subgraph(dag2, c("A", "B", "C", ancA, ancB, ancC))
```

```
  Random/Generated Bayesian network

  model:
   [A][B][C]
  nodes:                                3
  arcs:                                 0
     undirected arcs:                   0
     directed arcs:                     0
  average markov blanket size:          0.00
  average neighbourhood size:           0.00
  average branching factor:             0.00

  generation algorithm:                 Empty
```

As there is no path between A and B in this graph, A and B are d-separated given C.

6.4 Consider the survey data set from Chapter 1.

1. Learn a BN with the IAMB algorithm and the asymptotic mutual information test.

2. Learn a second BN with IAMB but using only the first 100 observations of the data set. Is there a significant loss of information in the resulting BN compared to the BN learned from the whole data set?

3. Repeat the structure learning in the previous point with IAMB and the Monte Carlo and sequential Monte Carlo mutual information tests. How do the resulting networks compare with the BN learned with the asymptotic test? Is the increased execution time justified?

1.
```
survey <- read.table("../discrete/survey.txt", header = TRUE,
            colClasses = "factor")
dag <- iamb(survey, test = "mi")
```

2.
```
dag100 <- iamb(survey[1:100, ], test = "mi")
nrow(directed.arcs(dag))
  [1] 0
nrow(undirected.arcs(dag))
  [1] 8
nrow(directed.arcs(dag100))
  [1] 0
```

```
nrow(undirected.arcs(dag100))
```
```
[1] 2
```

While both DAGs are very different from that in Figure 1.1, dag100 has only
a single arc; not enough information is present in the first 100 observations
to learn the correct structure. In both cases all arcs are undirected. After
assigning directions with cextend, we can see that dag100 has a much lower
score than dag, which confirms that dag100 is not as good a fit for the data
as dag.

```
score(cextend(dag), survey, type = "bic")
```
```
[1] -2000.152
```
```
score(cextend(dag100), survey, type = "bic")
```
```
[1] -2008.116
```

3. The BIC score computed from the first 100 observations does not increase
   when using Monte Carlo tests, and the DAGs we learn still have just a
   single arc. There is no apparent benefit over the corresponding asymptotic
   test.

```
dag100.mc <- iamb(survey[1:100, ], test = "mc-mi")
narcs(dag100.mc)
```
```
[1] 1
```
```
dag100.smc <- iamb(survey[1:100, ], test = "smc-mi")
narcs(dag100.smc)
```
```
[1] 1
```
```
score(cextend(dag100.mc), survey, type = "bic")
```
```
[1] -2008.116
```
```
score(cextend(dag100.smc), survey, type = "bic")
```
```
[1] -2008.116
```

6.5 Consider again the survey data set from Chapter 1.

1. **Learn a BN using Bayesian posteriors for both structure and
   parameter learning, in both cases with iss = 5.**

2. **Repeat structure learning with hc and 3 random restarts and with
   tabu. How do the BNs differ? Is there any evidence of numerical
   or convergence problems?**

3. **Use increasingly large subsets of the survey data to check
   empirically that BIC and BDe are asymptotically equivalent.**

1. ```
   dag <- hc(survey, score = "bde", iss = 5)
   bn <- bn.fit(dag, survey, method = "bayes", iss = 5)
   ```

2. ```
   dag.hc3 <- hc(survey, score = "bde", iss = 5, restart = 3)
   dag.tabu <- tabu(survey, score = "bde", iss = 5)
   modelstring(dag.hc3)
   ```
   ```
   [1] "[O][S][E|O:S][A|E][R|E][T|R]"
   ```
   ```
   modelstring(dag.tabu)
   ```
   ```
   [1] "[O][S][E|O:S][A|E][R|E][T|R]"
   ```

The two DAGs are quite different; from the model strings above, hc seems to learn a structure that is closer to that in Figure 1.1. The BIC scores of dag.hc3 and dag.tabu support the conclusion that hc with random restarts is a better fit for the data.

```
score(dag.hc3, survey)
 [1] -1999.733
score(dag.tabu, survey)
 [1] -1999.733
```

Using the debug option to explore the learning process we can confirm that no numerical problem is apparent, because all the DAGs learned from the random restarts fit the data reasonably well.

```
3.  for (size in c(0.005, 0.01, 0.1, 0.5, 1)) {
       subset = sample(nrow(survey), nrow(survey) * size)
       subsample = survey[subset, ]
       print(score(dag.hc3, subsample, type = "bic") /
             score(dag.hc3, subsample, type = "bde"))
    }
    [1] 0.8072155
    [1] 1.138477
    [1] 1.032488
    [1] 1.007157
    [1] 1.003957
```

The ratio between the two scores is more indicative than difference: their values are not standardised hence the absolute magnitude of the difference is meaningless.

6.6 Consider the marks data set from Section 6.7.

  1. **Create a bn object describing the graph in the bottom-right panel of Figure 6.7 and call it mdag.**
  2. **Construct the skeleton, the CPDAG and the moral graph of mdag.**
  3. **Discretise the marks data using "interval" discretisation with 2, 3 and 4 intervals.**
  4. **Perform structure learning with hc on each of the discretised data sets; how do the resulting DAGs differ?**

```
1.  mdag <- model2network(paste("[ANL][MECH][LAT|ANL:MECH]",
            "[VECT|LAT][ALG|LAT][STAT|LAT]", sep = ""))
```

```
2.  mdag.sk <- skeleton(mdag)
    mdag.cpdag <- cpdag(mdag)
    mdag.moral <- moral(mdag)
```

```
3.  data(marks)
    dmarks2 <- discretize(marks, "interval", breaks = 2)
    dmarks3 <- discretize(marks, "interval", breaks = 3)
    dmarks4 <- discretize(marks, "interval", breaks = 4)
```

4.
```
dag2 <- hc(dmarks2)
dag3 <- hc(dmarks3)
dag4 <- hc(dmarks4)
modelstring(dag2)
  [1] "[MECH][VECT|MECH][ALG|VECT][ANL|ALG][STAT|ALG]"
modelstring(dag3)
  [1] "[MECH][ALG|MECH][ANL|ALG][STAT|ALG][VECT|ANL]"
modelstring(dag4)
  [1] "[MECH][VECT][ALG][ANL|ALG][STAT|ANL]"
```

From the output above, we can deduce that using 4 intervals for discretising the data breaks all the dependencies between the variables; dag4 has only two arcs. Using 2 or 3 intervals results in DAGs with 4 arcs, which is closer to the 6 arcs of the true structure. However, the DAGs are still quite different from the latter, suggesting that a noticeable amount of information is lost in the discretisation.

Bibliography

Agresti, A. (2013). *Categorical Data Analysis*. Wiley, 3rd edition.

Aliferis, C. F., Statnikov, A., Tsamardinos, I., Mani, S., and Xenofon, X. D. (2010). Local Causal and Markov Blanket Induction for Causal Discovery and Feature Selection for Classification Part I: Algorithms and Empirical Evaluation. *Journal of Machine Learning Research*, 11:171–234.

Andersen, S., Olesen, K., Jensen, F., and Jensen, F. (1989). HUGIN – a Shell for Building Bayesian Belief Universes for Expert Systems. In *Proceedings of the 11th International Joint Conference on Artificial Intelligence (IJCAI)*, pages 1080–1085. Morgan Kaufmann Publishers Inc., San Francisco, CA, USA.

Anderson, T. W. (2003). *An Introduction to Multivariate Statistical Analysis*. Wiley, 3rd edition.

Andreassen, S., Jensen, F. V., Andersen, S. K., Falck, B., Kjærulff, U., Woldbye, M., Sørensen, A. R., Rosenfalck, A., and Jensen, F. (1989). MUNIN – An Expert EMG Assistant. In Desmedt, J. E., editor, *Computer-Aided Electromyography and Expert Systems*, pages 255–277. Elsevier.

Ash, R. B. (2000). *Probability and Measure Theory*. Academic Press, 2nd edition.

Balov, N. and Salzman, P. (2020). *catnet: Categorical Bayesian Network Inference*.

Bang-Jensen, J. and Gutin, G. (2009). *Digraphs: Theory, Algorithms and Applications*. Springer, 2nd edition.

Bøttcher, S. G. and Dethlefsen, C. (2003). deal: A Package for Learning Bayesian Networks. *Journal of Statistical Software*, 8(20):1–40.

Bouckaert, R. R. (1995). *Bayesian Belief Networks: From Construction to Inference*. PhD thesis, Utrecht University, The Netherlands.

Carpenter, B., Gelman, A., Hoffman, M. D., Lee, D., Goodrich, B., Betancourt, M., Brubaker, M., Guo, J., Li, P., and Riddell, A. (2017). Stan: A Probabilistic Programming Language. *Journal of Statistical Software*, 76(1):1–32.

Carpenter, B., Hoffman, M. D., Brubaker, M., Lee, D., Li P., and Betancourt, M. (2015). *The Stan Math Library: Reverse-Mode Automatic Differentiation*.

Castillo, E., Gutiérrez, J. M., and Hadi, A. S. (1997). *Expert Systems and Probabilistic Network Models*. Springer.

Centers for Disease Control and Prevention (2004). *The 1999–2004 Dual Energy X-ray Absorptiometry (DXA) Multiple Imputation Data Files and Technical Documentation*. National Center for Health Statistics, Hyattsville, MD (USA).

Centers for Disease Control and Prevention (2010). *National Health and Nutrition Examination Survey: Body Composition Procedures Manual*. National Center for Health Statistics, Hyattsville, MD (USA).

Chao, Y.-S., Wu, H.-T., Scutari, M., Chen, T.-S., Wu, C.-J., Durand, M., and Boivin, A. (2017). A Network Perspective on Patient Experiences and Health Status: The Medical Expenditure Panel Survey 2004 to 2011. *BMC Health Services Research*, 17(579):1–12.

Cheng, J. and Druzdzel, M. J. (2000). AIS-BN: An Adaptive Importance Sampling Algorithm for Evidential Reasoning in Large Bayesian Networks. *Journal of Artificial Intelligence Research*, 13:155–188.

Cooper, G. F. and Yoo, C. (1999). Causal Discovery from a Mixture of Experimental and Observational Data. In *Proceedings of the 15th Annual Conference on Uncertainty in Artificial Intelligence*, pages 116–125. Morgan Kaufmann, San Francisco, CA, USA.

Crawley, M. J. (2013). *The R Book*. Wiley, 2nd edition.

Cussens, J. (2012). Bayesian Network Learning with Cutting Planes. In *Proceedings of the 27th Conference on Uncertainty in Artificial Intelligence*, pages 153–160.

DeGroot, M. H. and Scherivsh, M. J. (2012). *Probability and Statistics*. Prentice Hall, 4th edition.

Denis, J.-B. (2020). *rbmn: Handling Linear Gaussian Bayesian Networks*.

Diestel, R. (2005). *Graph Theory*. Springer, 3rd edition.

Edwards, D. I. (2000). *Introduction to Graphical Modelling*. Springer, 2nd edition.

Eggeling, R., Viinikka, J., Vuoksenmaa, A., and Koivisto, M. (2019). On Structure Priors for Learning Bayesian Networks. In *Proceedings of the 22nd International Conference on Artificial Intelligence and Statistics*, pages 1687–1695.

Feller, W. (1968). *An Introduction to Probability Theory and Its Applications*. Wiley, 3rd edition.

Friedman, N. and Koller, D. (2003). Being Bayesian about Bayesian Network Structure: A Bayesian Approach to Structure Discovery in Bayesian Networks. *Machine Learning*, 50(1–2):95–126.

Friedman, N., Linial, M., Nachman, I., and Pe'er, D. (2000). Using Bayesian Network to Analyze Expression Data. *Journal of Computational Biology*, 7:601–620.

Friedman, N., Pe'er, D., and Nachman, I. (1999). Learning Bayesian Network Structure from Massive Datasets: The "Sparse Candidate" Algorithm. In *Proceedings of 15th Conference on Uncertainty in Artificial Intelligence*, pages 206–221. Morgan Kaufmann, San Francisco, CA, USA.

Gasse, M., Aussem, A., and Elghazel, H. (2014). A Hybrid Algorithm for Bayesian Network Structure Learning with Application to Multi-Label Learning. *Expert Systems with Applications*, 41(15):6755–6772.

Goeman, J. J. (2018). *penalized R Package*.

Hartemink, A. J. (2001). *Principled Computational Methods for the Validation and Discovery of Genetic Regulatory Networks*. PhD thesis, School of Electrical Engineering and Computer Science, Massachusetts Institute of Technology.

Hartigan, J. A. (1975). *Clustering Algorithms*. Wiley.

Hartley, S. W. and Sebastiani, P. (2013). PleioGRiP: Genetic Risk Prediction with Pleiotropy. *Bioinformatics*, 29(8):1086–1088.

Hastie, T., Tibshirani, R., and Friedman, J. (2009). *The Elements of Statistical Learning: Data Mining, Inference, and Prediction*. Springer, 2nd edition.

Hausser, J. and Strimmer, K. (2009). Entropy Inference and the James-Stein Estimator, with Application to Nonlinear Gene Association Networks. *Journal of Machine Learning Research*, 10:1469–1484.

Heckerman, D., Geiger, D., and Chickering, D. M. (1995). Learning Bayesian Networks: The Combination of Knowledge and Statistical Data. *Machine Learning*, 20(3):197–243. Available as Technical Report MSR-TR-94-09.

Højsgaard, S. (2020). *gRain: Graphical Independence Networks*.

Højsgaard, S., Dethlefsen, C., and Bowsher, C. (2020). *gRbase: A Package for Graphical Modelling in R*.

Højsgaard, S., Edwards, D., and Lauritzen, S. (2012). *Graphical Models with R*. Use R! series. Springer.

Holmes, D. E. and Jaim, L. C. (2008). *Innovations in Bayesian Networks: Theory and Applications*. Springer.

Ide, J. S. and Cozman, F. G. (2002). Random Generation of Bayesian Networks. In *Proceedings of the 16th Brazilian Symposium on Artificial Intelligence (SBIA)*, pages 366–375. Springer-Verlag.

Johnson, N. L., Kemp, A. W., and Kotz, S. (2005). *Univariate Discrete Distributions*. Wiley, 3rd edition.

Johnson, N. L., Kotz, S., and Balakrishnan, N. (1994). *Continuous Univariate Distributions*, volume 1. Wiley, 2nd edition.

Johnson, N. L., Kotz, S., and Balakrishnan, N. (1995). *Continuous Univariate Distributions*, volume 2. Wiley, 2nd edition.

Johnson, N. L., Kotz, S., and Balakrishnan, N. (1997). *Discrete Multivariate Distributions*. Wiley.

Kalisch, M., Mächler, M., Colombo, D., Maathuis, M. H., and Bühlmann, P. (2012). Causal Inference Using Graphical Models with the R Package pcalg. *Journal of Statistical Software*, 47(11):1–26.

Kenett, R. S., Perruca, G., and Salini, S. (2012). *Modern Analysis of Customer Surveys: With Applications Using R*, chapter 11. Wiley.

Kjærluff, U. B. and Madsen, A. L. (2013). *Bayesian Networks and Influence Diagrams: A Guide to Construction and Analysis*. Springer, 2nd edition.

Koivisto, M. and Sood, K. (2004). Exact Bayesian Structure Discovery in Bayesian Networks. *Journal of Machine Learning Research*, 5:549–573.

Koller, D. and Friedman, N. (2009). *Probabilistic Graphical Models: Principles and Techniques*. MIT Press.

Korb, K. B. and Nicholson, A. E. (2011). *Bayesian Artificial Intelligence*. Chapman and Hall, 2nd edition.

Koski, T. and Noble, J. M. (2009). *Bayesian Networks: An Introduction*. Wiley.

Kotz, S., Balakrishnan, N., and Johnson, N. L. (2000). *Continuous Multivariate Distributions*. Wiley, 2nd edition.

Kratzer, G. and Furrer, R. (2019). *mcmcabn: a Structural MCMC Sampler for DAGs Learned from Observed Systemic Datasets*.

Kratzer, G., Pittavino, M., Lewis, F. I., and Furrer, R. (2019). *abn: an R Package for Modelling Multivariate Data Using Additive Bayesian Networks*.

Kruschke, J. K. (2014). *Doing Bayesian Data Analysis: A Tutorial with R, JAGS, and Stan*. Academic Press.

Kucukelbir, A., Ranganath, R., Gelman, A., and Blei, D. (2015). Automatic Variational Inference in Stan. In *Advances in Neural Information Processing Systems 28*, pages 568–576.

Kuipers, J. and Moffa, G. (2017). Partition MCMC for Inference on Acyclic Digraphs. *Journal of the American Statistical Association*, 112(517):282–299.

Kuipers, J., Moffa, G., and Heckerman, D. (2014). Addendum on the Scoring of Gaussian Directed Acyclic Graphical Models. *The Annals of Statistics*, 42(4):1689–1691.

Kulinskaya, E., Morgenthaler, S., and Staudte, R. G. (2008). *Meta Analysis: A Guide to Calibrating and Combining Statistical Evidence*. Wiley.

Kullback, S. (1968). *Information Theory and Statistics*. Dover Publications.

Larrañaga, P., Sierra, B., Gallego, M. J., Michelena, M. J., and Picaza, J. M. (1997). Learning Bayesian Networks by Genetic Algorithms: A Case Study in the Prediction of Survival in Malignant Skin Melanoma. In *Proceedings of the 6th Conference on Artificial Intelligence in Medicine in Europe (AIME'97)*, pages 261–272. Springer, Berlin.

Lauritzen, S. (1996). *Graphical Models*. Oxford University Press.

Loève, M. (1977). *Probability Theory*. Springer-Verlag, 4th edition.

Lunn, D., Spiegelhalter, D., Thomas, A., and Best, N. (2009). The BUGS Project: Evolution, Critique and Future Directions (with discussion). *Statistics in Medicine*, 28(5):3049–3067.

Lunn, D., Thomas, A., Best, N., and Spiegelhalter, D. (2000). WinBUGS–A Bayesian Modelling Framework: Concepts, Structure, and Extensibility. *Statistics and Computing*, 10(4):325–337.

Mahalanobis, P. C. (1936). On the Generalised Distance in Statistics. *Proceedings of the National Institute of Sciences of India*, 2(1):49–55.

Mardia, K. V., Kent, J. T., and Bibby, J. M. (1979). *Multivariate Analysis*. Academic Press.

Margaritis, D. (2003). *Learning Bayesian Network Model Structure from Data*. PhD thesis, School of Computer Science, Carnegie-Mellon University, Pittsburgh, PA. Available as Technical Report CMU-CS-03-153.

McElreath, R. (2016). *Statistical Rethinking: A Bayesian Course with Examples in R and Stan*. Chapman and Hall.

Melançon, G., Dutour, I., and Bousquet-Mélou, M. (2001). Random Generation of Directed Acyclic Graphs. *Electronic Notes in Discrete Mathematics*, 10:202–207.

Morota, G., Valente, B. D., Rosa, G. J. M., Weigel, K. A., and Gianola, D. (2012). An Assessment of Linkage Disequilibrium in Holstein Cattle Using a Bayesian Network. *Journal of Animal Breeding and Genetics*, 129(6):474–487.

Murphy, K. P. (2012). *Machine Learning: A Probabilistic Perspective*. MIT Press.

Nagarajan, R., Scutari, M., and Lèbre, S. (2013). *Bayesian Networks in R with Applications in Systems Biology*. Use R! series. Springer.

Neapolitan, R. E. (2003). *Learning Bayesian Networks*. Prentice Hall.

Pearl, J. (1988). *Probabilistic Reasoning in Intelligent Systems: Networks of Plausible Inference*. Morgan Kaufmann.

Pearl, J. (2009). *Causality: Models, Reasoning and Inference*. Cambridge University Press, 2nd edition.

Pinheiro, J. C. and Bates, D. M. (2000). *Mixed-Effects Models in S and S-PLUS*. Springer.

Plummer, M. (2003). JAGS: A Program for Analysis of Bayesian Graphical Models Using Gibbs Sampling. In *Proceedings of the 3rd International Workshop on Distributed Statistical Computing (DSC)*, pages 1–10.

Plummer, M., Best, N., Cowles, K., and Vines, K. (2006). CODA: Convergence Diagnosis and Output Analysis for MCMC. *R News*, 6(1):7–11.

Pourret, O., Naim, P., and Marcot, B. (2008). *Bayesian Networks: A Practical Guide to Applications*. Wiley.

Robert, C. P. and Casella, G. (2009). *Introducing Monte Carlo Methods with R*. Springer.

Ross, S. (2018). *A First Course in Probability*. Prentice Hall, 10th edition.

Russell, S. J. and Norvig, P. (2009). *Artificial Intelligence: A Modern Approach*. Prentice Hall, 3rd edition.

Sachs, K., Perez, O., Pe'er, D., Lauffenburger, D. A., and Nolan, G. P. (2005). Causal Protein-Signaling Networks Derived from Multiparameter Single-Cell Data. *Science*, 308(5721):523–529.

Schäfer, J., Opgen-Rhein, R., Zuber, V., Ahdesmäki, M., Silva, A. P. D., and Strimmer., K. (2017). *corpcor: Efficient Estimation of Covariance and (Partial) Correlation*.

Scutari, M. (2010). Learning Bayesian Networks with the bnlearn R Package. *Journal of Statistical Software*, 35(3):1–22.

Scutari, M. (2016). An Empirical-Bayes Score for Discrete Bayesian Networks. *Journal of Machine Learning Research (Proceedings Track, PGM 2016)*, 52:438–448.

Scutari, M. (2020). Bayesian Network Models for Incomplete and Dynamic Data. *Statistica Neerlandica*. In print.

Scutari, M. and Brogini, A. (2012). Bayesian Network Structure Learning with Permutation Tests. *Communications in Statistics – Theory and Methods*, 41(16–17):3233–3243.

Scutari, M., Graafland, C. E., and Gutiérrez, J. M. (2019). Who Learns Better Bayesian Network Structures: Accuracy and Speed of Structure Learning Algorithms. *International Journal of Approximate Reasoning*, 115:235–253.

Scutari, M. and Nagarajan, R. (2013). On Identifying Significant Edges in Graphical Models of Molecular Networks. *Artificial Intelligence in Medicine*, 57(3):207–217.

Shäfer, J. and Strimmer, K. (2005). A Shrinkage Approach to Large-Scale Covariance Matrix Estimation and Implications for Functional Genomics. *Statistical Applications in Genetics and Molecular Biology*, 4:32.

Sinoquet, C. and Mourad, R. (2013). *Probabilistic Graphical Models for Genetics, Genomics, and Postgenomics*. Oxford University Press.

Spector, P. (2009). *Data Manipulation with R*. Springer-Verlag.

Spirtes, P., Glymour, C., and Scheines, R. (2000). *Causation, Prediction, and Search*. MIT Press, 2nd edition.

Stan Development Team (2020a). *PyStan: the Python interface to Stan*.

Stan Development Team (2020b). *RStan: the R interface to Stan*.

Sucar, L. E. (2015). *Probabilistic Graphical Models: Principles and Applications*. Springer.

Suzuki, J. (2017). An Efficient Bayesian Network Structure Learning Strategy. *New Generation Computing*, 35(1):105–124.

Tsamardinos, I., Aliferis, C. F., and Statnikov, A. (2003). Algorithms for Large Scale Markov Blanket Discovery. In *Proceedings of the 16th International Florida Artificial Intelligence Research Society Conference*, pages 376–381. AAAI Press.

Tsamardinos, I. and Borboudakis, G. (2010). Permutation Testing Improves Bayesian Network Learning. In Balcázar, J., Bonchi, F., Gionis, A., and Sebag, M., editors, *Machine Learning and Knowledge Discovery in Databases*, pages 322–337. Springer.

Tsamardinos, I., Brown, L. E., and Aliferis, C. F. (2006). The Max-Min Hill-Climbing Bayesian Network Structure Learning Algorithm. *Machine Learning*, 65(1):31–78.

Ueno, M. (2011). Robust Learning of Bayesian Networks for Prior Belief. In *Proceedings of the 27th Conference on Uncertainty in Artificial Intelligence*, pages 698–707.

Venables, W. N. and Ripley, B. D. (2002). *Modern Applied Statistics with S*. Springer, 4th edition.

Verma, T. S. and Pearl, J. (1991). Equivalence and Synthesis of Causal Models. *Uncertainty in Artificial Intelligence*, 6:255–268.

Villa-Vialaneix, N., Liaubet, L., Laurent, T., Cherel, P., Gamot, A., and SanCristobal, M. (2013). The Structure of a Gene Co-Expression Network Reveals Biological Functions Underlying eQTLs. *PLOS ONE*, 8(4):e60045.

Wang, K.-J., Chen, J.-L., and Wang, K.-M. (2019). Medical Expenditure Estimation by Bayesian Network for Lung Cancer Patients at Different Severity Stages. *Computers in Biology and Medicine*, 106:97–105.

Whittaker, J. (1990). *Graphical Models in Applied Multivariate Statistics*. Wiley.

Yaramakala, S. and Margaritis, D. (2005). Speculative Markov Blanket Discovery for Optimal Feature Selection. In *Proceedings of the 5th IEEE International Conference on Data Mining*, pages 809–812. IEEE Computer Society, Los Alamitos, CA.

Yu, J., Smith, V. A., Wang, P. P., Hartemink, A. J., and Jarvis, E. D. (2004). Advances to Bayesian Network Inference for Generating Causal Networks from Observational Biological Data. *Bioinformatics*, 20(18):3594–3603.

Yuan, C. and Druzdzel, M. J. (2003). An Importance Sampling Algorithm Based on Evidence Pre-Propagation. In *Proceedings of the 19th Conference on Uncertainty in Artificial Intelligence*, pages 624–631.

Zou, M. and Conzen, S. D. (2005). A New Dynamic Bayesian Network (DBN) Approach for Identifying Gene Regulatory Networks from Time Course Microarray Data. *Bioinformatics*, 21(1):71–79.

Index

Printed in the United States
by Baker & Taylor Publisher Services

Printed in the United States
by Baker & Taylor Publisher Services